Teaching Mathematics
in the
Visible Learning Classroom

Grades 3–5

Teaching Mathematics in the Visible Learning Classroom

Grades 3–5

John Almarode, Douglas Fisher,
Kateri Thunder, Sara Delano Moore,
John Hattie, and Nancy Frey

CORWIN Mathematics

FOR INFORMATION:

Corwin

A SAGE Company

2455 Teller Road

Thousand Oaks, California 91320

(800) 233-9936

www.corwin.com

SAGE Publications Ltd.

1 Oliver's Yard

55 City Road

London EC1Y 1SP

United Kingdom

SAGE Publications India Pvt. Ltd.

B 1/I 1 Mohan Cooperative Industrial Area

Mathura Road, New Delhi 110 044

India

SAGE Publications Asia-Pacific Pte Ltd

18 Cross Street #10-10/11/12

China Square Central

Singapore 048423

Executive Editor, Mathematics: Erin Null

Editorial Development Manager: Julie Nemer

Senior Editorial Assistant: Jessica Vidal

Production Editor: Tori Mirsadjadi

Copy Editor: Christina West

Typesetter: C&M Digitals (P) Ltd.

Proofreader: Susan Schon

Indexer: Laurie Andriot

Cover Designer: Rose Storey

Marketing Manager: Margaret O'Connor

Copyright © 2019 by Corwin

Printed in the United States of America

Library of Congress Cataloging-in-Publication Data

Names: Almarode, John, author.

Title: Teaching mathematics in the visible learning classroom, grades 3-5 / John Almarode [and five others].

Description: Thousand Oaks, California : Corwin, a Sage Company, [2019] | Includes bibliographical references and index.

Identifiers: LCCN 2018046544 | ISBN 9781544333243 (pbk. : alk. paper)

Subjects: LCSH: Mathematics teachers—In-service training. | Mathematics—Study and teaching (Elementary)

Classification: LCC QA10.5 .T433 2019 | DDC 372.7/044—dc23
LC record available at https://lccn.loc.gov/2018046544

This book is printed on acid-free paper.

SUSTAINABLE FORESTRY INITIATIVE

Certified Chain of Custody
Promoting Sustainable Forestry
www.sfiprogram.org
SFI-01268

SFI label applies to text stock

19 20 21 22 23 10 9 8 7 6 5 4 3 2 1

Contents

List of Videos

Chapter 4. Teaching for Procedural Knowledge and Fluency

Chapter 5. Knowing Your Impact: Evaluating for Mastery

Note From the Publisher: The authors have provided video and web content throughout the book that is available to you through QR (quick response) codes. To read a QR code, you must have a smartphone or tablet with a camera. We recommend that you download a QR code reader app that is made specifically for your phone or tablet brand.

 online resources

Videos may also be accessed at **resources.corwin.com/ vlmathematics-3-5**

Acknowledgments

We are forever grateful for the teachers and instructional leaders who strive each and every day to make an impact in the lives of learners. Their dedication to teaching and learning is evident in the video clips linked to the QR codes in this book. The teachers in Charlottesville, Virginia, have graciously opened their classrooms and conversations to us, allowing us to make mathematics in the Visible Learning classroom visible to readers. The learners they work with in the Charlottesville City Public Schools are better simply because they spent time with the following people:

Mrs. Jenny Isaacs-Lowe, Special Educator, Venable Elementary School

Mr. Christopher Lorigan, Third Grade Teacher, Burnley-Moran Elementary School

Ms. Isabel Smith, Fourth Grade Teacher, Burnley-Moran Elementary School

Mrs. Rachel Caldwell, Fourth Grade Teacher, Burnley-Moran Elementary School

Mrs. Calder McLellan, Mathematics Specialist, Burnley-Moran Elementary School

Mr. James Henderson, Assistant Superintendent, Charlottesville City Schools

We are extremely grateful to Superintendent Dr. Rosa Atkins for allowing us into the schools and classrooms of Charlottesville, helping to make our work come alive.

Ms. Christen Showker is an excellent teacher in Rockingham County Public Schools in Virginia. Ms. Beth Buchholz, Ms. Hollins Mills, and Ms. Katy Campbell are excellent teachers in public schools. They are actively engaged in implementing Visible Learning into their classrooms. Their contributions to this book provide clear examples of how they have taken the Visible Learning research and translated the findings into their teaching and learning. We are forever grateful to these four teachers for sharing their journey with us so that we could share these examples with you.

About the Authors

John Almarode, PhD, has worked with schools, classrooms, and teachers all over the world. John began his career in Augusta County, Virginia, teaching mathematics and science to a wide range of students. In addition to spending his time in preK–12 schools and classrooms, he is an associate professor in the Department of Early, Elementary, and Reading Education and the codirector of James Madison University's Center for STEM Education and Outreach. In 2015, John was named the Sarah Miller Luck Endowed Professor of Education. However, what really sustains John—and what marks his greatest accomplishment—is his family. John lives in Waynesboro, Virginia, with his wife, Danielle, a fellow educator; their two children, Tessa and Jackson; and their Labrador Retrievers, Angel and Forest. John can be reached at www.johnalmarode.com.

Douglas Fisher, PhD, is Professor of Educational Leadership at San Diego State University and a teacher leader at Health Sciences High & Middle College. He is the recipient of a William S. Grey Citation of Merit and NCTE's Farmer Award for Excellence in Writing, as well as a Christa McAuliffe Award for Excellence in Teacher Education. Doug can be reached at dfisher@mail.sdsu.edu.

Kateri Thunder, PhD, served as an inclusive, early childhood educator, an Upward Bound educator, a mathematics specialist, an assistant professor of mathematics education at James Madison University, and site director for the Central Virginia Writing Project (a National Writing Project site at the University of Virginia). Kateri is a member of the Writing Across the Curriculum Research Team with Dr. Jane Hansen, co-author of *The Promise of Qualitative Metasynthesis for Mathematics Education*, and co-creator of *The Math Diet*. Currently, Kateri has followed her passion back to the classroom. She teaches in an at-risk preK program, serves as the PreK–4 Math Lead for Charlottesville City Schools, and works as an educational consultant. Kateri is happiest exploring the world with her best friend and husband, Adam, and her family. Kateri can be reached at www.mathplusliteracy.com.

Sara Delano Moore, PhD, is Director of Professional Learning at ORIGO Education. A fourth-generation educator, her work focuses on helping teachers and students understand mathematics as a coherent and connected discipline through the power of deep understanding and multiple representations for learning. Sara has worked as a classroom teacher of mathematics and science in the elementary and middle grades, a mathematics teacher educator, Director of the Center for Middle School Academic Achievement for the Commonwealth of Kentucky, and Director of Mathematics & Science at ETA hand2mind. Her journal articles appear in *Mathematics Teaching in the Middle School*, *Teaching Children Mathematics*, *Science & Children*, and *Science Scope*. Sara can be reached at sara@sdmlearning.com.

John Hattie, PhD, has been Laureate Professor of Education and Director of the Melbourne Education Research Institute at the University of Melbourne, Australia, since March 2011. He was previously Professor of Education at the University of Auckland, as well as in North Carolina, Western Australia, and New England. His research interests are based on applying measurement models to education problems. He has been president of the International Test Commission, has served as adviser to various ministers, chairs the Australian Institute for Teachers and School Leaders, and in the 2011 Queen's Birthday Honours was made "Order of Merit for New Zealand" for his services to education. He is a cricket umpire and coach, enjoys being a dad to his young men, is besotted with his dogs, and moved with his wife as she attained a promotion to Melbourne. Learn more about his research at www.corwin.com/visiblelearning.

Nancy Frey, PhD, is Professor of Literacy in the Department of Educational Leadership at San Diego State University. She is the recipient of the 2008 Early Career Achievement Award from the National Reading Conference and is a teacher leader at Health Sciences High & Middle College. She is also a credentialed special educator, reading specialist, and administrator in California.

Introduction

Dylan is a precocious fourth grader who loves mathematics. One of his favorite pastimes is playing the 24 Game (Suntex International Inc., 1988). For those of us not familiar with this particular game, Dylan will quickly show you that this competitive game involves a card containing four numbers (e.g., 7, 5, 4, and 3). Once the card is placed on the table, each player in the game tries to figure out how to make the number 24 using addition, subtraction, multiplication, and division. For the example with 7, 5, 4, and 3, Dylan gave the following answer:

$$7 - 5 = 2$$

$$4 \times 3 = 12$$

$$12 \times 2 = 24$$

In this specific example, Dylan rattled off the difference between seven and five, the product of four and three, and multiplied those two answers to get 24. To note, Dylan was able to solve this particular problem before the teacher had finished placing the card on the table.

Dylan demonstrates a high level of proficiency, or mastery, in procedural knowledge in the area of computation involving the four basic operations with single-digit whole numbers (e.g., additive thinking and multiplicative thinking). However, there is more to Dylan's mathematics learning than his mastery of number facts. Dylan possesses a balance of conceptual understanding, procedural knowledge, and the ability to apply those concepts and thinking skills to a variety of mathematical contexts. By balance, we mean that no one dimension of mathematics learning is more important than the other two. Conceptual understanding, procedural knowledge, and the application of concepts

Teaching Takeaway

Procedural knowledge comes from balanced mathematics teaching and learning.

and thinking skills are each essential aspects of learning mathematics. Dylan's prowess in the 24 Game is not the result of his teachers implementing procedural knowledge, conceptual understanding, and application in isolation, but through a series of linked learning experiences and challenging mathematical tasks that result in him engaging in both mathematical content and processes.

If you were to engage in a conversation with Dylan about mathematics, you would quickly see that he is able to discuss the concept of multiplication and describe different ways to represent multiplication (i.e., equal groups, arrays, and number line models). Furthermore, he can articulate which model he prefers and why: "I sometimes pick the model based on the type of problem. You know, some ways work better with certain problems." Dylan also recognizes that he must apply this conceptual understanding and thinking to solving problems involving rates and price. He says, "If a pencil from the school store costs 10 cents and I want to buy five pencils, I need 50 cents." Dylan also mentions that he could easily use this information when he learns about geometric measurements next year. "Well, that is what my teacher tells me," he adds. Dylan's mathematics learning is not by chance, but by design. His progression in conceptual understanding, procedural knowledge, and the application of concepts and thinking skills come from the purposeful, deliberate, and intentional decisions of his current and past teachers. These decisions focus on the following:

- *What works best* and *what works best when* in the teaching and learning of mathematics, and
- Building and supporting assessment-capable visible learners in mathematics.

Our Learning Intention: To understand what works best in the mathematics classroom, Grades 3–5.

This book explores the components in mathematics teaching and learning in Grades 3–5 through the lens of *what works best* in student learning at the surface, deep, and transfer phases. We fully acknowledge that not every student in your classroom is like Dylan. Our students come to our classrooms with different background knowledge, levels of readiness, and learning needs. Our goal is to unveil what works best so that your learners develop the tools needed for successful mathematics learning.

What Works Best

Identifying what works best draws from the key findings from Visible Learning (Hattie, 2009) and also guides the classrooms described in this book. One of those key findings is that *there is no one way to teach mathematics or one best instructional strategy that works in all situations for all students*, but there is compelling evidence for certain strategies and approaches that have a greater likelihood of helping students reach their learning goals. In this book, we use the effect size information that John Hattie has collected and analyzed over many years to inform how we transform the findings from the Visible Learning research into learning experiences and challenging mathematical tasks that are most likely to have the strongest influence on student learning.

For readers less familiar with Visible Learning, we would like to take a moment to review what we mean by *what works best*. The Visible Learning database is composed of over 1,800 meta-analyses of studies that include over 80,000 studies and 300 million students. Some have argued that it is the largest educational research database amassed to date. To make sense of so much data, John Hattie focused his work on meta-analyses. A **meta-analysis** is a statistical tool for combining findings from different studies, with the goal of identifying patterns that can inform practice. In other words, a meta-analysis is a study of studies. The mathematical tool that aggregates the information is an effect size and can be represented by Cohen's *d*. An **effect size** is the magnitude, or relative size, of a given effect. Effect size information helps readers understand not only that something does or does not have an influence on learning but also the relative impact of that influence.

For example, imagine a hypothetical study in which pausing instruction to engage in a quick exercise or "brain break" results in relatively higher mathematics scores among fourth graders. Schools and classrooms around the country might feel compelled to devote significant time and energy to the development and implementation of brain breaks in all fourth grade classrooms in a specific district. However, let's say the results of this hypothetical study also indicate that the use of brain breaks had an effect size of 0.02 in mathematics achievement over the control group, an effect size pretty close to zero. Furthermore, the large number of students participating in the study made it almost certain

A **meta-analysis** is a statistical tool for combining findings from different studies, with the goal of identifying patterns that can inform practice.

Effect size represents the magnitude of the impact that a given approach has.

Video 1
What Is Visible Learning
for Mathematics?

To read a QR code, you must
have a smartphone or tablet with
a camera. We recommend that
you download a QR code reader
app that is made specifically for
your phone or tablet brand.

Videos can also be accessed at
*https://resources.corwin.com/
vlmathematics-3-5*

there would be a difference in the two groups of students (those participating in brain breaks versus those not participating in brain breaks). As an administrator or teacher, would you still devote large amounts of professional learning and instructional time on brain breaks? How confident would you be in the impact or influence of your decision on mathematics achievement in your district or school?

This is where an effect size of 0.02 for the "brain breaks effect" is helpful in discerning what works best in mathematics teaching and learning. Understanding the effect size helps us know how powerful a given influence is in changing achievement—in other words, the impact for the effort or return on the investment. The effect size helps us understand not just what works, but *what works best*. With the increased frequency and intensity of mathematics initiatives, programs, and packaged curricula, deciphering where to best invest resources and time to achieve the greatest learning outcomes for all students is challenging and frustrating. For example, some programs or packaged curricula are hard to implement and have very little impact on student learning, whereas others are easy to implement but still have limited influence on student growth and achievement in mathematics. This is, of course, on top of a literacy program, science kits, and other demands on the time and energy of elementary school teachers. Teaching mathematics in the Visible Learning classroom involves searching for those things that have the greatest impact and produce the greatest gains in learning, some of which will be harder to implement and some of which will be easier to implement.

As we begin planning for our unit on rational numbers, knowing the effect size of different influences, strategies, actions, and approaches to teaching and learning proves helpful in deciding where to devote our planning time and resources. Is a particular approach (e.g., classroom discussion, exit tickets, use of calculators, a jigsaw activity, computer-assisted instruction, simulation creation, cooperative learning, instructional technology, presentation of clear success criteria, development of a rubric, etc.) worth the effort for the desired learning outcomes of that day, week, or unit? With the average effect size across all influences measuring 0.40, John Hattie was able to demonstrate that influences, strategies, actions, and approaches with an effect size greater than 0.40 allow students to learn at an appropriate rate, meaning at least a year of growth for a year in school. Effect sizes greater than 0.40 mean

THE BAROMETER OF INFLUENCE

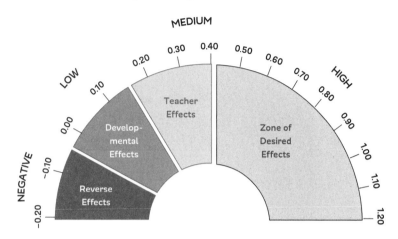

Source: Adapted from Hattie, J. (2009). *Visible learning: A synthesis of over 800 meta-analyses relating to achievement.* Figure 2.4, page 19. New York, NY: Routledge.

Figure I.1

more than a year of growth for a year in school. Figure I.1 provides a visual representation of the range of effect sizes calculated in the Visible Learning research.

Before this level was established, teachers and researchers did not have a way to determine an acceptable threshold, and thus we continued to use weak practices, often supported by studies with statistically significant findings.

Consider the following examples. First, let us consider classroom discussion or the use of mathematical discourse (see NCTM, 1991). Should teachers devote resources and time into planning for the facilitation of classroom discussion? Will this approach to mathematics provide a return on investment rather than "chalk talk," where we work out lots of problems on the board for students to include in their notes? With classroom discussion, teachers intentionally design and purposefully plan for learners to talk with their peers about specific problems or approaches to problems (e.g., comparing and contrasting strategies for multiplying and dividing large numbers versus small numbers,

EFFECT SIZE FOR ABILITY GROUPING (TRACKING/ STREAMING) = 0.12

THE BAROMETER FOR THE INFLUENCE OF CLASSROOM DISCUSSION

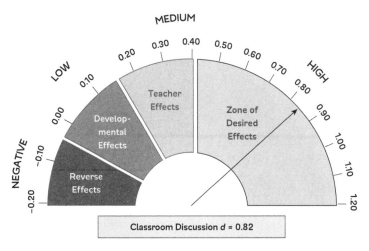

Source: Adapted from Hattie, J. (2009). *Visible learning: A synthesis of over 800 meta-analyses relating to achievement.* Figure 2.4, page 19. New York, NY: Routledge.

Figure I.2

Ability grouping, also referred to as tracking or streaming, is the long-term grouping or tracking of learners based on their ability. This is different from flexibly grouping students to work on a specific concept, skill, or application or address a misconception.

explaining their development of a formula for a three-dimensional shape) in collaborative groups. Peer groups might engage in working to solve complex problems or tasks (e.g., determining the equivalent decimal for a fraction using a number line). Although they are working in collaborative groups, the students would not be **ability grouped**. Instead, the teacher purposefully groups learners to ensure that there is academic diversity in each group as well as language support and varying degrees of interest and motivation. As can be seen in the barometer in Figure I.2, the effect size of classroom discussion is 0.82, which is well above our threshold and is likely to accelerate learning gains.

Therefore, individuals teaching mathematics in the Visible Learning classroom would use classroom discussions to understand mathematics learning through the eyes of their students and for students to see themselves as their own mathematic teachers.

Second, let us look at the use of calculators. Within academic circles, teacher workrooms, school hallways, and classrooms, there have been

THE BAROMETER FOR THE INFLUENCE OF USING CALCULATORS

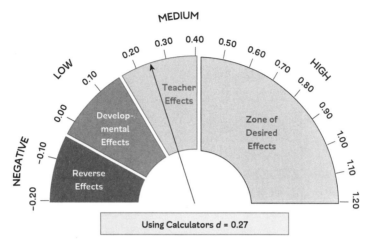

Source: Adapted from Hattie, J. (2009). *Visible learning: A synthesis of over 800 meta-analyses relating to achievement.* Figure 2.4, page 19. New York, NY: Routledge.

Figure I.3

many conversations about the use of the calculator in mathematics. There have been many efforts to reduce the reliance on calculators while at the same time developing technology-enhanced items on assessments in mathematics. Using a barometer as a visual representation of effect sizes, we see that the use of calculators has an overall effect size of 0.27. The barometer for the use of calculators is in Figure I.3.

As you can see, the effect size of 0.27 is below the zone of desired effects of 0.40. The evidence suggests that the impact of the use of calculators on mathematics achievement is low. However, closer examination of the five meta-analyses and the 222 studies that produced an overall effect size of 0.27 reveals a deeper story to the use of calculators. Calculators are most effective in the following circumstances: (1) when they are used for computation, deliberate practice, and learners checking their work; (2) when they are used to reduce the amount of cognitive load on learners as they engage in problem solving; and (3) when there is an intention behind using them (e.g., generating a pattern of square numbers, computing multiples of 10, or calculating the area or volume

EFFECT SIZE FOR CLASSROOM DISCUSSION = **0.82**

EFFECT SIZE OF USE OF CALCULATORS = **0.27**

of a large space or object). This leads us into a second key finding from John Hattie's Visible Learning research: *We should not hold any influence, instructional strategy, action, or approach to teaching and learning in higher esteem than students' learning.*

What Works Best When

Visible Learning in the mathematics classroom is a continual evaluation of our impact on student learning. From the above example, the use of calculators is not really the issue and should not be our focus. Instead, our focus should be on the intended learning outcomes for that day and how calculators support that learning. Visible Learning is more than a checklist of dos and don'ts. Rather than checking influences with high effect sizes off the list and scratching out influences with low effect sizes, we should match the *best* strategy, action, or approach with learning needs of our students. In other words, is the use of calculators the right strategy or approach for the learners at the right time, for this specific content? Clarity about the learning intention brings into focus what the learning is for the day, why students are learning about this particular piece of content and process, and how we *and* our learners will know they have learned the content. Teaching mathematics in the Visible Learning classroom is not about a specific strategy, but a location in the learning process.

Visible Learning in the mathematics classroom occurs when teachers *see* learning through the eyes of their students and students *see* themselves as their own teachers. How do teachers of mathematics see multiplicative thinking, rational numbers, and geometric measurements through the eyes of their students? In turn, how do teachers develop assessment-capable visible learners—students who see themselves as their own teachers—in the study of numbers, operations, and relationships? Mathematics teaching and learning, where *teachers see learning through the eyes of their learners and learners see themselves as their own teachers*, results from specific, intentional, and purposeful decisions about each of these dimensions of mathematics instruction critical for student growth and achievement. Conceptualizing, implementing, and sustaining Visible Learning in the mathematics classroom by identifying *what works best* and *what works best when* is exactly what we set out to do in this book.

Over the next several chapters, we will show how to support mathematics learners in their pursuit of conceptual understanding, procedural knowledge, and application of concepts and thinking skills through the lens of *what works best when*. This requires us, as mathematics teachers, to be clear in our planning and preparation for each learning experience and challenging mathematics tasks. Using the guiding questions in Figure I.4, we will model how to blend what works best with what works best *when*. You can use these questions in your own planning. This planning guide is found also in Appendix B.

Through these specific, intentional, and purposeful decisions in our mathematics instruction, we pave the way for helping learners see themselves as their own teachers, thus making them assessment-capable visible learners in mathematics.

The Path to Assessment-Capable Visible Learners in Mathematics

Teaching mathematics in the Visible Learning classroom builds and supports assessment-capable visible learners (Frey, Hattie, & Fisher, 2018). With an effect size of 1.33, providing a mathematics learning environment that allows learners to see themselves as their own teacher is essential in today's classrooms.

Ava is a bubbly fourth grader who loves school. She loves school for all of the right reasons—learning and socializing. At times, she confuses the two, but she quickly engages in the day's mathematics lesson. During her review, Ava is engaging in the deliberate practice of adding fractions with unlike denominators. This is a topic that is challenging to her and is important background or prior knowledge for upcoming learning. During a discussion with her shoulder partner, Ava discusses her areas of strength and areas for growth: "I am good at adding fractions when the bottom numbers—wait, the denominators—are the same. You know, you just add the top numbers. I need more practice when the number—I mean, the denominator—is different. I have to slow down and figure it out." This is a characteristic of an assessment-capable learner in mathematics.

EFFECT SIZE FOR ASSESSMENT-CAPABLE VISIBLE LEARNERS = **1.33**

Video 2
Creating Assessment-Capable Visible Learners

https://resources.corwin.com/vlmathematics-3-5

PLANNING FOR CLARITY GUIDE

I have to be clear about what content and practice or process standards I am using to plan for clarity. Am I using only mathematics standards or am I integrating other content standards (e.g., writing, reading, or science)?

Rather than what I want my students to be doing, this question focuses on the learning. What do the standards say my students should learn? The answer to this question generates the **learning intentions** for this particular content.

Once I have clear learning intentions, I must decide when and how to communicate them with my learners. Where does it best fit in the instructional block to introduce the day's learning intentions? Am I going to use guiding questions?

As I gather evidence about my students' learning progress, I need to establish what they should know, understand, and be able to do that would demonstrate to me that they have learned the content. This list of evidence generates the **success criteria** for the learning.

ESTABLISHING PURPOSE

1 What are the key content standards I will focus on in this lesson?
Content Standards:

2 What are the learning intentions (the goal and *why* of learning, stated in student-friendly language) I will focus on in this lesson?
Content:
Language:
Social:

3 When will I introduce and reinforce the learning intention(s) so that students understand it, see the relevance, connect it to previous learning, and can clearly communicate it themselves?

SUCCESS CRITERIA

4 What evidence shows that students have mastered the learning intention(s)? What criteria will I use?

I can statements:

online resources — This planning guide is available for download at **resources.corwin.com/vlmathematics-3-5.**

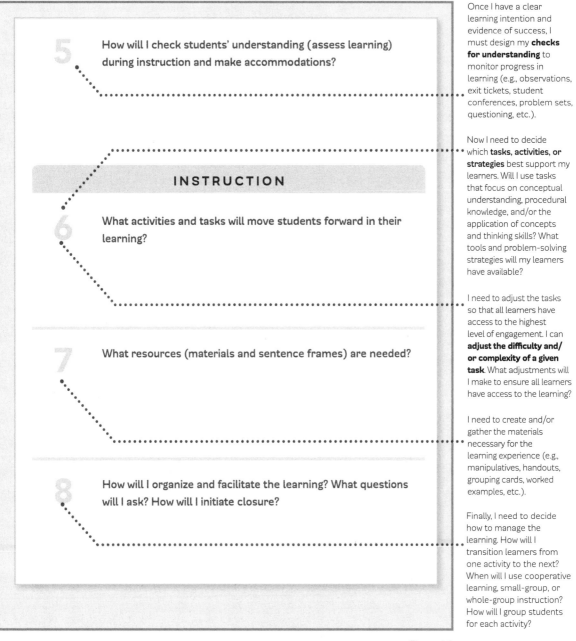

5   How will I check students' understanding (assess learning) during instruction and make accommodations?

Once I have a clear learning intention and evidence of success, I must design my **checks for understanding** to monitor progress in learning (e.g., observations, exit tickets, student conferences, problem sets, questioning, etc.).

INSTRUCTION

6   What activities and tasks will move students forward in their learning?

Now I need to decide which **tasks, activities, or strategies** best support my learners. Will I use tasks that focus on conceptual understanding, procedural knowledge, and/or the application of concepts and thinking skills? What tools and problem-solving strategies will my learners have available?

7   What resources (materials and sentence frames) are needed?

I need to adjust the tasks so that all learners have access to the highest level of engagement. I can **adjust the difficulty and/ or complexity of a given task**. What adjustments will I make to ensure all learners have access to the learning?

I need to create and/or gather the materials necessary for the learning experience (e.g., manipulatives, handouts, grouping cards, worked examples, etc.).

8   How will I organize and facilitate the learning? What questions will I ask? How will I initiate closure?

Finally, I need to decide how to manage the learning. How will I transition learners from one activity to the next? When will I use cooperative learning, small-group, or whole-group instruction? How will I group students for each activity?

Figure I.4

Assessment-capable visible mathematics learners are:

1. Active in their mathematics learning. Learners deliberately and intentionally engage in learning mathematics content and processes by asking themselves questions, monitoring their own learning, and taking the reins of their learning. They know their current level of learning.

Later in the lesson, Ava is working in a cooperative learning group on finding the area of the school garden. Although the concept of area is a review, her teacher is using a concept Ava is familiar with to add context to two- by two-digit multiplication. Her cooperative learning group has encountered a challenging calculation, 27 × 16. However, they quickly recognize that they have the tools to solve this problem. One of the group members chimes in, "To find the product, the answer to the problem, 27 × 16, I am going to use an open array model. These numbers are unfriendly." This is a characteristic of an assessment-capable learner in mathematics.

Assessment-capable visible mathematics learners are:

2. Able to plan the immediate next steps in their mathematics learning within a given unit of study or topic. Because of the active role taken by an assessment-capable visible mathematics learner, these students can plan their next steps and select the right tools (e.g., manipulatives, problem-solving approaches, and/or metacognitive strategies) to guide their learning. They know what additional tools they need to successfully move forward in a task or topic.

Ava's teacher, Ms. Christen Showker, takes time to individually conference with each student at least once a week. This allows the teacher to provide very specific feedback on each learner's progress. Ava begins the conference by stating, "Yesterday's exit ticket surprised me. You [Ms. Showker] wrote on my paper that I needed to revisit place value. I think I mixed up the thousands place. So, tomorrow I am going to work out the entire process for finding which number is larger in my notebook and not try and do it all in my head." This is a characteristic of an assessment-capable learner in mathematics.

Assessment-capable visible mathematics learners are:

3. Aware of the purpose of the assessment and feedback provided by peers and the teacher. Whether the assessment is informal, formal, formative, or summative, assessment-capable visible mathematics learners have a firm understanding of the information behind each assessment and the feedback exchanged in the classroom. Put differently, these learners not only seek feedback, but they recognize that errors are opportunities for learning, monitor their progress, and adjust their learning (adapted from Frey et al., 2018) (see Figure I.5).

Over the next several chapters, we will explore how to create a classroom environment that focuses on learning and provides the best environment for developing assessment-capable visible mathematics learners who can engage in the mathematical habits of mind represented in one form or another in every standards document. Such learners can achieve the following:

1. Make sense of problems and persevere in solving them.

2. Reason abstractly and quantitatively.

3. Construct viable arguments and critique the reasoning of others.

4. Model with mathematics.

5. Use appropriate tools strategically.

6. Attend to precision.

7. Look for and make use of structure.

8. Look for and express regularity in repeated reasoning (© Copyright 2010. National Governors Association Center for Best Practices and Council of Chief State School Officers. All rights reserved.).

How This Book Works

As authors, we assume you have read *Visible Learning for Mathematics* (Hattie et al., 2017), so we are not going to recount all of the information

ASSESSMENT-CAPABLE LEARNERS:

 KNOW THEIR CURRENT LEVEL OF UNDERSTANDING

 KNOW WHERE THEY'RE GOING AND ARE CONFIDENT TO TAKE ON THE CHALLENGE

 SELECT TOOLS TO GUIDE THEIR LEARNING

 SEEK FEEDBACK AND RECOGNIZE THAT ERRORS ARE OPPORTUNITIES TO LEARN

 MONITOR THEIR PROGRESS AND ADJUST THEIR LEARNING

 RECOGNIZE THEIR LEARNING AND TEACH OTHERS

Source: Adapted from Frey, Hattie, & Fisher (2018).

Figure I.5

contained in that book. Rather, we are going to dive deeper into aspects of mathematics instruction in Grades 3–5 that are critical for students' success, helping you to envision what a Visible Learning mathematics classroom like yours looks like. In each chapter, we profile three teachers who have worked to make mathematics learning visible for their students and have influenced learning in significant ways. Each chapter will do the following:

1. Provide effect sizes for specific influences, strategies, actions, and approaches to teaching and learning.

2. Provide support for specific strategies and approaches to teaching mathematics.

3. Incorporate content-specific examples from third, fourth, and fifth grade mathematics curricula.

4. Highlight aspects of assessment-capable visible learners.

Through the eyes of third, fourth, and fifth grade mathematics teachers, as well as the additional teachers and the instructional leaders in the accompanying videos, we aim to show you the mix and match of strategies you can use to orchestrate your lessons in order to help your students build their conceptual understanding, procedural knowledge, and application of concepts and thinking skills in the most visible ways possible—visible to you and to them. If you are a mathematics specialist, mathematics coordinator, or methods instructor, you may be interested in exploring the vertical progression of these content areas across preK–12 within Visible Learning classrooms and see how visible learners grow and progress across time and content areas. Although you may identify with one of the teachers from a content perspective, we encourage you to read all of the vignettes to get a full sense of the variety of choices you can make in your instruction, based on your instructional goals.

In Chapter 1, we focus on the aspects of mathematics instruction that must be included in each lesson. We explore the components of effective mathematics instruction (conceptual, procedural, and application) and note that there is a need to recognize that student learning has to occur at the surface, deep, and transfer levels within each of these components. Surface, deep, and transfer learning served as the organizing feature of *Visible Learning for Mathematics*, and we will briefly review them and their value in learning. This book focuses on the ways in which teachers can develop students' surface, deep, and transfer learning, specifically by supporting students, conceptual understanding, procedural knowledge, and application whether with comparing fractions or geometric measurement. Finally, Chapter 1 contains information about the use of checks for understanding to monitor student learning. Generating evidence of learning is important for both teachers and students in determining the impact of the learning experiences and challenging mathematical tasks on learning. If learning is not happening, then we must make adjustments.

Following this introductory chapter, we turn our attention, separately, to each component of mathematics teaching and learning. However, we will walk through the process starting with the application of concepts and thinking skills, then direct our attention to conceptual understanding, and finally, procedural knowledge. This seemingly unconventional approach will allow us to start by making the goal or endgame visible: learners applying mathematics concepts and thinking skills to other situations or contexts.

Chapter 2 focuses on *application* of concepts and thinking skills. Returning to our three profiled classrooms, we will look at how we plan, develop, and implement challenging mathematical tasks that scaffold student thinking as they apply their learning to new contexts or situations. Teaching mathematics in the Visible Learning classroom means supporting learners as they use mathematics in a variety of situations. In order for learners to effectively apply mathematical concepts and thinking skills to different situations, they must have strong conceptual understanding and procedural knowledge. Returning to Figure I.4, we will walk through the process for establishing clear learning intentions, defining evidence of learning, and developing challenging tasks that, as you have already come to expect, encourage learners to see themselves as their own teachers. Each chapter will discuss how to differentiate mathematical tasks by adjusting their difficulty and/or complexity, working to meet the needs of all learners in the mathematics classroom.

Chapters 3 and 4 take a similar approach with conceptual understanding and procedural knowledge, respectively. Using Chapter 2 as a reference point, we will return to the three profiled classrooms and explore the conceptual understanding and procedural knowledge that provided the foundation for their learners applying ideas to different mathematical situations. For example, what influences, strategies, actions, and approaches support a learner's conceptual understanding of multiplication and division, rational numbers, or geometric measurement? With conceptual understanding, what works best as we encourage learners to see mathematics as more than a set of mnemonics and procedures? Supporting students' thinking as they focus on underlying conceptual principles and properties, rather than relying on memory cues like PEMDAS, also necessitates adjusting the difficulty and complexity of mathematics tasks. As in Chapter 2, we will talk about differentiating tasks by adjusting their difficulty and complexity.

In this book, we do not want to discourage the value of procedural knowledge. Although mathematics is more than procedural knowledge, developing skills in basic procedures is needed for later work in each area of mathematics from the area and circumference of a circle to linear equations. As in the previous two chapters, Chapter 4 will look at what works best when supporting students' procedural knowledge. Adjusting the difficulty and complexity of tasks will once again help us meet the needs of all learners.

In the final chapter of this book, we focus on how to make mathematics learning visible through evaluation. Teachers must have clear knowledge of their impact so that they can adjust the learning environment. Learners must have clear knowledge about their own learning so that they can be active in the learning process, plan the next steps, and understand what is behind the assessment. What does evaluation look like so that teachers can use it to plan instruction and to determine the impact that they have on learning? As part of Chapter 5, we highlight the value of feedback and explore the ways in which teachers can provide effective feedback to students that is growth producing. Furthermore, we will highlight how learners can engage in self-regulation feedback and provide feedback to their peers.

This book contains information on critical aspects of mathematics instruction in Grades 3–5 that have evidence for their ability to influence student learning. We're not suggesting that these be implemented in isolation, but rather that they be combined into a series of linked learning experiences that result in students engaging in mathematics learning more fully and deliberately than they did before. Whether finding equivalent fractions or calculating volume, we strive to create a mathematics classroom where we *see* learning through the eyes of our students and students *see* themselves as their own mathematics teachers. As learners progress from simplifying rational expressions to using ratios and proportions, teaching mathematics in the Visible Learning classroom should build and support assessment-capable visible mathematics learners.

Please allow us to introduce you to Christen Showker, Beth Buchholz, Hollins Mills, and Katy Campbell. These four elementary school teachers set out each day to deliberately, intentionally, and purposefully impact the mathematics learning of their students. Whether they teach third, fourth, or fifth grade, they recognize that:

- They have the capacity to select and implement various teaching and learning strategies that enhance their students' learning in mathematics.

- The decisions they make about their teaching have an impact on students' learning.

- Each student can learn mathematics, and they need to take responsibility to teach all learners.

- They must continuously question and monitor the impact of their teaching on student learning. (adapted from Hattie & Zierer, 2018)

Mindframes are ways of thinking about teaching and learning. Teachers who possess certain ways of thinking have major impacts on student learning.

Through the videos accompanying this book, you will meet additional elementary teachers and the instructional leaders who support them in their teaching. Collectively, the recognitions above—or their **mindframes**—lead to action in their mathematics classrooms and their actions lead to outcomes in student learning. This is where we begin our journey through *Teaching Mathematics in the Visible Learning Classroom*.

TEACHING WITH CLARITY IN MATHEMATICS

1

CHAPTER 1 SUCCESS CRITERIA:

(1) I can describe teacher clarity and the process for providing clarity in my classroom.

(2) I can describe the components of effective mathematics instruction.

(3) I can relate the learning process to my own teaching and learning.

(4) I can give examples of how to differentiate mathematics tasks.

(5) I can describe the four different approaches to teaching mathematics.

A **learning intention** describes what it is that we want our students to learn.

Success criteria specify the necessary evidence students will produce to show their progress toward the learning intention.

EFFECT SIZE FOR LEARNING INTENTION = 0.68

EFFECT SIZE FOR SUCCESS CRITERIA = 1.13

EFFECT SIZE FOR COOPERATIVE LEARNING = 0.40

EFFECT SIZE FOR COOPERATIVE LEARNING COMPARED TO COMPETITIVE LEARNING = 0.53

In Ms. Showker's fourth grade mathematics class, students are learning to collect, organize, and represent data using line graphs or bar graphs. Ms. Showker starts the math block by walking her learners through the **learning intention** and **success criteria**.

Learning Intention: I am learning that the type of data and the way I display that data are connected.

Success Criteria:

1. I can describe why I would use a graph.

2. I can compare and contrast a line graph with a bar graph.

3. I can explain why I would use each type of graph.

4. I can construct a line graph and a bar graph from data.

There are many different approaches for engaging learners in data, line graphs, and bar graphs. Given that the specific standard associated with today's learning emphasizes questions and investigations related to students' experiences, interests, and environment, Ms. Showker uses data collected during their unit on weather and an earlier unit on measurement. During these two units, Ms. Showker's learners collected weather data (i.e., sky cover and precipitation type) and kept those observations in their interactive notebooks. She introduces today's lesson as follows:

> Over the past several weeks, we have recorded weather observations in your interactive notebooks. We used tally marks to record the sky cover for the day (for example, cloudy or sunny). We also used tally marks to record the type of precipitation (for example, rain, snow, or none). When I say "go," please get out your interactive math notebooks and make your way to your assigned tables.

Learners are flexibly grouped based on the previous day's exit ticket. There are learners who demonstrate surface knowledge about the necessary parts of graphs (e.g., axes, labels, and key). Furthermore, previous checks for understanding provided evidence about her learners' understanding and use of skip counting, an important piece of prior knowledge for this particular content. Ms. Showker provides each table with a folder of resources that will support students in accomplishing today's mathematics task. Each folder contains several examples of line graphs and several examples of bar graphs.

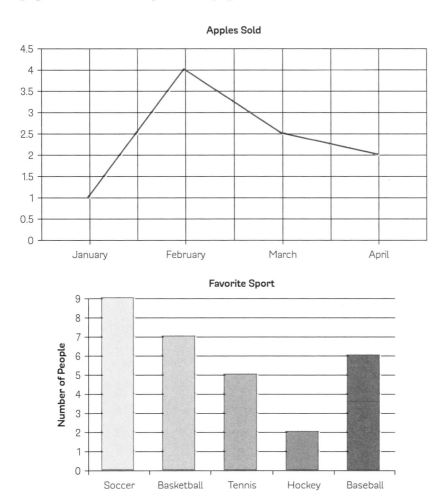

Ms. Showker deliberately informs her learners that their folders contain examples of two types of graphs: line graphs and bar graphs. She does not want vocabulary or terminology to distract from today's learning. She says, "With your fellow mathematicians, please sort the examples into two groups—line graphs and bar graphs." As she monitors her learners during this sorting task, Ms. Showker is making note of the specific mathematical discussions around the sorting of the

EFFECT SIZE
FOR DIRECT/
DELIBERATE
INSTRUCTION
= 0.60

examples. She notices some students are focusing on essential characteristics (e.g., bar graphs contain vertical or horizontal bars with a separate bar for each category, or line graphs include a key to identify what each line represents), whereas others are basing their sorts on irrelevant characteristics (e.g., this graph is about sports, and this one is not about sports). Ms. Showker uses the evidence gathered during these discussions to provide direct/deliberate instruction through a mini-lesson for specific learners who need additional instruction to master the concept.

The challenging task during today's mathematics block is for each group to construct a line graph and a bar graph using the weather data in their interactive notebooks. Learners have to first decide which type of graph best fits the two different types of data (i.e., type of sky cover and precipitation). They will then create graphs, using the examples in the folder as a model. Anticipating that learners would be at different places in the learning progression associated with data and graphs, Ms. Showker prepared different levels of support for each group. In addition to the examples provided for the sorting task, she provides groups the following different levels of support:

EFFECT SIZE
FOR FINDING THE
"RIGHT" LEVEL OF
CHALLENGE = 0.74

- Graph paper with rows or columns drawn for each category, and options for values of pictures
- Stamps or stickers to use in constructing the graphs
- Graph paper with the axes drawn and scaled (learners need to graph the data, label the axes, provide a key if necessary, and add a title)
- Graph paper with the axes drawn, but not scaled
- Graph paper with a checklist of the components needed for each type of graph
- A blank sheet of paper

EFFECT SIZE FOR
SCAFFOLDING
= 0.82

EFFECT SIZE FOR
QUESTIONING
= 0.48

Throughout the task, Ms. Showker monitors her learners' progress, asking guiding questions and providing feedback and additional support as needed. She wants to give her learners an opportunity for productive struggle, but she carefully monitors this struggle to ensure her students do not get frustrated.

Before Ms. Showker collects the graphs from each group, she asks them to complete an individual writing prompt.

> As you wrap up today's task, I want you to summarize your learning on the left side of your interactive mathematics notebook by responding to the following writing prompt: What informed your decisions about how to best represent each type of data?

Ms. Showker is implementing the principles of Visible Learning in her fourth grade mathematics classroom. Our intention is to help you implement these principles in your own classroom. By providing learners with a challenging task, a clear learning intention and success criteria, and direct/deliberate instruction where and when needed, Ms. Showker's cooperative learning teams are developing conceptual understanding, gaining procedural knowledge, and applying their learning. She holds high expectations for her students in terms of both the difficulty and complexity of the task, as well as her learners' ability to deepen their mathematics learning by making learning visible to herself and each individual learner. As Ms. Showker monitors the learning progress in each team, holding all students individually accountable for their own learning, she takes opportunities to provide additional instruction when needed. Although her learners are engaged in cooperative learning with their peers, she regularly assesses her students to identify gaps in their learning that she can address with additional instruction or intervention. Ms. Showker is mobilizing principles of Visible Learning through her conscious awareness of her impact on student learning, and her students are consciously aware of their learning through challenge tasks. Ms. Showker works to accomplish this through these specific, intentional, and purposeful decisions in her mathematics instruction. She had clarity in her mathematics teaching, allowing her learners to have clarity and see themselves as their own teachers (i.e., assessment-capable visible mathematics learners). This came about from using the following guiding questions in her planning and preparation for learning:

1. What do I want my students to learn?

2. What evidence shows that the learners have mastered the learning or are moving toward mastery?

Video 3
What Does Teacher Clarity Mean in Grades 3–5 Mathematics?

https://resources.corwin.com/ vlmathematics-3-5

EFFECT SIZE FOR
TEACHER CLARITY
= 0.75

HOW VISIBLE TEACHING AND VISIBLE LEARNING COMPARE

Visible Teaching	Visible Learning
Clearly communicates the learning intention	Understands the intention of the learning experience
Identifies challenging success criteria	Knows what success looks like
Uses a range of learning strategies	Develops a range of learning strategies
Continually monitors student learning	Knows when there is no progress and makes adjustments
Provides feedback to learners	Seeks feedback about learning

Figure 1.1

online resources ➤ This figure is available for download at **resources.corwin.com/vlmathematics-3-5**.

3. How will I check learners' understanding and progress?

4. What tasks will get my students to mastery?

5. How will I differentiate tasks to meet the needs of all learners?

6. What resources do I need?

7. How will I manage the learning?

Ms. Showker exemplifies the relationship between Visible Teaching and Visible Learning (see Figure 1.1).

Now, let's look at how to achieve **clarity** in teaching mathematics by first understanding how components of mathematics learning interface with the learning progressions of the students in our classrooms. Then, we will use this understanding to establish learning intentions, identify success criteria, create challenging mathematical tasks, and monitor or check for understanding.

> **Clarity** in learning means that both the teacher and the student know what the learning is for the day, why they are learning it, and what success looks like.

Components of Effective Mathematics Learning

Mathematics is more than just memorizing formulas and then working problems with those formulas. Rather than using a compilation of procedures (memorizing mnemonics for the different types of polygons, labeling place value, or multiplying two whole numbers), mathematics learning involves an interplay of conceptual understanding, procedural knowledge, and application of mathematical concepts and thinking skills. Together these compose rigorous mathematics learning, which is furthered by the Standards for Mathematical Practice that claim students should:

1. Make sense of problems and persevere in solving them.

2. Reason abstractly and quantitatively.

3. Construct viable arguments and critique the reasoning of others.

4. Model with mathematics.

5. Use appropriate tools strategically.

6. Attend to precision.

7. Look for and make use of structure.

8. Look for and express regularity in repeated reasoning (© Copyright 2010. National Governors Association Center for Best Practices and Council of Chief State School Officers. All rights reserved).

Teaching mathematics in the Visible Learning classroom fosters student growth through attending to these mathematical practices or processes. As highlighted by Ms. Showker in the opening of this chapter, this comes from linked learning experiences and challenging mathematics tasks that make learning visible to both students and teachers.

Surface, Deep, and Transfer Learning

Each school year, regardless of the grade level, students develop their mathematics prowess through a progression that moves from understanding the surface contours of a concept into how to work with that concept efficiently by leveraging procedural skills as well as applying concepts and thinking skills to an ever-deepening exploration of what lies beneath mathematical ideas. For example, third graders transition from an emphasis on number sense involving whole numbers to a focus on decimals

THE RELATIONSHIP BETWEEN SURFACE, DEEP, AND TRANSFER LEARNING IN MATHEMATICS

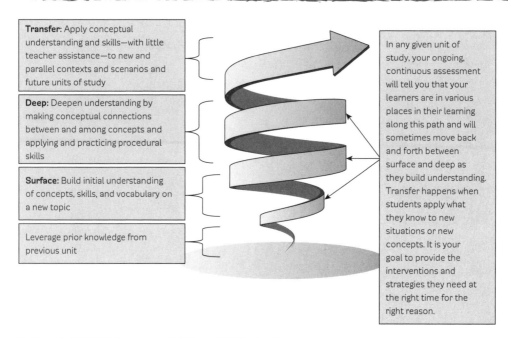

Transfer: Apply conceptual understanding and skills—with little teacher assistance—to new and parallel contexts and scenarios and future units of study

Deep: Deepen understanding by making conceptual connections between and among concepts and applying and practicing procedural skills

Surface: Build initial understanding of concepts, skills, and vocabulary on a new topic

Leverage prior knowledge from previous unit

In any given unit of study, your ongoing, continuous assessment will tell you that your learners are in various places in their learning along this path and will sometimes move back and forth between surface and deep as they build understanding. Transfer happens when students apply what they know to new situations or new concepts. It is your goal to provide the interventions and strategies they need at the right time for the right reason.

Source: Hattie et al. (2017). Spiral Image copyright EssentialsCollection/iStock.com

Figure 1.2

Surface learning is the phase in which students build initial conceptual understanding of a mathematical idea and learn related vocabulary, representations, and procedural skills.

and fractions. As another example, learners progress in their mathematics learning from third to fifth grade through an increased emphasis on using different representations of numbers to engage in problem solving. Understanding these progressions requires that teachers consider the levels of learning expected from students. We think of three levels, or phases, of learning: surface, deep, and transfer (see Figure 1.2).

Learning is a process, not an event. With some conceptual understanding, procedural knowledge, and application, students may still only understand at the surface level. We do not define surface-level learning as superficial learning. Rather, we define **surface learning** as the initial development of conceptual understanding and procedural skill, with some application. In other words, this is the students' initial, often foundational, learning around what a fraction is, the various representations of fractions (e.g., region or area model, set models, or length models), and fundamental ideas about how to use fractions to solve problems. Surface learning is often misrepresented as rote rehearsal or memorization and is therefore not valued, but it is an essential part of the mathematics learning process. Students must understand how to represent fractions with manipulatives, in words or sketches, in context, and in real-world applications to be able to connect these representations and use them in an authentic situation.

With the purposeful and intentional use of learning strategies that focus on how to relate and extend ideas, surface mathematics learning becomes deep learning. **Deep learning** occurs when students begin to make *connections* among conceptual ideas and procedural knowledge and apply their thinking with greater fluency. As learners begin to monitor their progress, adjust their learning, and select strategies to guide their learning, they more efficiently and effectively plan, investigate, elaborate on their knowledge, and make generalizations based on their experiences with mathematics content and processes.

If learners are to deepen their knowledge, they must regularly encounter situations that foster the transfer and generalization of their learning. The American Psychological Association (2015) notes that "student transfer or generalization of their knowledge and skills is not spontaneous or automatic" (p. 10) and **transfer learning** requires intentionally created events on the part of the teacher.

Figure 1.3 contains a representative list of strategies or influences organized by phase of learning. This is an updated list from *Visible Learning for Mathematics* (Hattie et al., 2017). Notice how many of these strategies and influences—clarity of learning goals, questioning, discourse, and problem solving—align with the Effective Teaching Practices outlined by the National Council of Teachers of Mathematics (2014) in *Principles to Actions: Ensuring Mathematical Success for All* (see Figure 1.4).

For the influences from the Visible Learning research, we placed them in a specific phase based on the evidence of their impact and the outcomes that researchers use to document the impact each has on students' learning. For example, we have included concept maps and graphic organizers under deep learning. Learners will find it hard to organize

EFFECT SIZE FOR PRIOR ABILITY = **0.94**

EFFECT SIZE FOR PRIOR ACHIEVEMENT = **0.55**

Deep learning is a period when students consolidate their understanding and apply and extend some surface learning knowledge to support deeper conceptual understanding.

EFFECT SIZE FOR ELABORATION AND ORGANIZATION = **0.75**

Transfer learning is the point at which students take their consolidated knowledge and skills and apply what they know to new scenarios and different contexts. It is also a time when students are able to think more metacognitively, reflecting on their own learning and understanding.

HIGH-IMPACT APPROACHES AT EACH PHASE OF LEARNING

Surface Learning		Deep Learning		Transfer Learning	
Strategy	ES	Strategy	ES	Strategy	ES
Imagery	0.45	Inquiry-based teaching	0.40	Extended writing	0.44
Note taking	0.50	Questioning	0.48	Peer tutoring	0.53
Process skill: record keeping	0.52	Self-questioning	0.55	Synthesizing information across texts	0.63
Direct/deliberate instruction	0.60	Metacognitive strategy instruction	0.60	Problem-solving teaching	0.68
Organizing	0.60	Concept mapping	0.64	Formal discussions (e.g., debates)	0.82
Vocabulary programs	0.62	Reciprocal teaching	0.74	Organizing conceptual knowledge	0.85
Leveraging prior knowledge	0.65	Class discussion: discourse	0.82	Transforming conceptual knowledge	0.85
Mnemonics	0.76	Outlining and transforming notes	0.85	Identifying similarities and differences	1.32
Summarization	0.79	Small-group learning 0.47			
Integrating prior knowledge	0.93	Cooperative learning 0.40			
Teacher expectations 0.43					
Feedback 0.70					
Teacher clarity 0.75					
Integrated curricula programs 0.47					
Assessment-capable visible learner 1.33					

Source: Adapted from Almarode, Fisher, Frey, & Hattie (2018).

Figure 1.3

EFFECT SIZE FOR
METACOGNITIVE
STRATEGIES = 0.60
AND EVALUATION
AND REFLECTION
= 0.75

mathematics information or ideas visually or graphically if they do not yet understand that information. Without a conceptual understanding of the properties of the operations, fourth grade mathematics students may approach single-step and multistep problems based on surface-level features (e.g., this problem involves money or addition) instead of deep-level features (e.g., this problem requires me to use the distributive

EFFECTIVE MATHEMATICS TEACHING PRACTICES

Establish mathematics goals to focus learning. Effective teaching of mathematics establishes clear goals for the mathematics that students are learning, situates goals within learning progressions, and uses the goals to guide instructional decisions.

Implement tasks that promote reasoning and problem solving. Effective teaching of mathematics engages students in solving and discussing tasks that promote mathematical reasoning and problem solving and allow multiple entry points and varied solution strategies.

Use and connect mathematical representations. Effective teaching of mathematics engages students in making connections among mathematical representations to deepen understanding of mathematics concepts and procedures and as tools for problem solving.

Facilitate meaningful mathematical discourse. Effective teaching of mathematics facilitates discourse among students to build shared understanding of mathematical ideas by analyzing and comparing student approaches and arguments.

Pose purposeful questions. Effective teaching of mathematics uses purposeful questions to assess and advance students' reasoning and sense making about important mathematical ideas and relationships.

Build procedural fluency from conceptual understanding. Effective teaching of mathematics builds fluency with procedures on a foundation of conceptual understanding so that students, over time, become skillful in using procedures flexibly as they solve contextual and mathematical problems.

Support productive struggle in learning mathematics. Effective teaching of mathematics consistently provides students, individually and collectively, with opportunities and supports to engage in productive struggle as they grapple with mathematical ideas and relationships.

Elicit and use evidence of student thinking. Effective teaching of mathematics uses evidence of student thinking to assess progress toward mathematical understanding and to adjust instruction continually in ways that support and extend learning.

Source: NCTM. (2014). *Principles to actions: Ensuring mathematical success for all.* Reston, VA: NCTM, National Council of Teachers of Mathematics. Reprinted with permission.

Figure 1.4

property). When students have sufficient surface learning about specific content and processes, they are able to see the connections between multiple ideas and connect their specific knowledge of properties to analyze problems based on these deep-level features (i.e., the distributive property is applicable across multiple contexts), which allow for the generalization of mathematics principles. As a reminder, two key findings from the Visible Learning research are as follows:

1. There is no one way to teach mathematics or one best instructional strategy that works in all situations for all students; and

2. We should not hold any influence, instructional strategy, action, or approach in higher esteem than students' learning.

As teachers, our conversations should focus on identifying where students are in their learning journey and moving them forward in their learning. This is best accomplished by talking about learning and measuring the impact that various approaches have on students' learning. If a given approach is not working, change it. If you experienced success with a particular strategy or approach in the past, give it a try but make sure that the strategy or approach is working in this context. Just because we can use PEMDAS to support computation, for example, does not mean those mnemonics will work for all students in your mathematics classroom—particularly if they lack understanding of the conceptual underpinnings of those procedures. Teachers have to monitor the impact that learning strategies have on students' mathematics learning and how they are progressing from surface, to deep, to transfer.

> As teachers, our conversations should focus on identifying where students are in their learning journey and moving them forward in their learning.

Moving Learners Through the Phases of Learning

The **SOLO Taxonomy** (Structure of Observed Learning Outcomes) (Biggs & Collis, 1982) conceptualizes the movement from surface to deep to transfer learning as a process of first branching out and then strengthening connections between ideas (Figure 1.5).

As you reflect on your own students, you can likely think of learners who have limited to no prior experiences with certain formal mathematics content. They do, however, have significant informal prior knowledge. Take, for example, perimeter and area. Although learners have likely encountered real-world uses of these concepts (e.g., how many laps around the track equals a mile in physical education, a fence

> The **SOLO Taxonomy** is a framework that describes learners' thinking and understanding of mathematics. The taxonomy conceptualizes the learning process from surface, to deep, and then to transfer.

THE SOLO TAXONOMY

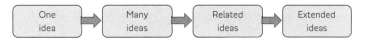

Source: Adapted from Biggs & Collis (1982).

Figure 1.5

around the garden or yard), many have had no experience with the formal mathematics behind those real-world applications. Thus, they have no formal, relevant structure to their thinking. This means they likely struggle to articulate a single idea about the perimeter or area of a given shape using mathematical language or notation.

Another example of this occurs with the equations or formulas for area and perimeter. Learners may recognize that letters represent specific items in an equation, say $A = l \times w$ or $P = l + l + w + w$ for a rectangle, but they are not able to identify these features in a rectangle or find the perimeter or area of a square when they are given only one side. This part of the SOLO Taxonomy is referred to as the prestructural level or prestructural thinking. At the prestructural level, learners may focus on irrelevant ideas, avoid engaging in the content, or not know where to start. In some cases, learners may ask for a ruler. This requires the teacher to support the learner in acquiring and building background knowledge. When teachers clearly recognize that a learner or learners are at the prestructural level, the learning experience should aim to build surface learning around concepts, procedures, and applications.

Surface Learning in the Intermediate Mathematics Classroom

As learners progress in their thinking, they may develop single ideas or a single aspect related to a concept. Learners at this level can identify and name shapes or attributes, follow simple procedures, highlight single aspects of a concept, and solve one type of problem (Hook & Mills, 2011). They know that $A = l \times w$ calculates the area of a rectangle

Teaching Takeaway

We must preassess our learners to identify their prior knowledge or background knowledge in the mathematics content they are learning. We should use informal language, little notation, and familiar contexts in our preassessments to allow all students to show what they know.

and that *l* represents the length and *w* represents the width. They can only solve problems involving the exact type of rectangle provided in an in-class example, such as in Figure 1.6.

EXAMPLES OF DIFFERENT AREA AND PERIMETER PROBLEMS INVOLVING QUADRILATERALS

Perimeter and Area

Find the area and perimeter area of the following squares and rectangles.

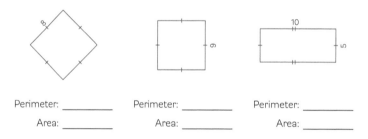

Perimeter: _____   Perimeter: _____   Perimeter: _____

Area: _____   Area: _____   Area: _____

Figure 1.6

For example, let's say a learner can calculate the area of a rectangle where the length and width are labeled on the diagram and the length is greater than the width. Any variation to the problem will pose a significant challenge to this learner, requiring additional instruction (e.g., the width is the larger number, the rectangle is rotated, or the dimensions are merely provided without a diagram). With the right approach at the right time, learners will continue to build surface learning by acquiring multiple ideas about concepts, procedures, and applications. Learners can then solve area problems involving different variations of rectangles or from different perspectives, and they describe coherently how to calculate the area of any rectangle instead of simply executing the algorithm. However, at this phase of their thinking and learning, learners see each variation of an area of a rectangle problem as a distinct scenario that is not connected to the other variations of rectangles.

Like Ms. Showker, all teachers should establish learning intentions and success criteria based on where students are in their learning progression. Moving away from perimeter and area and back to Ms. Showker's classroom, let us look at how we can develop learning intentions and success criteria for conceptual understanding, procedural knowledge, and application for learners at these two levels (one idea and many ideas) (Figures 1.7 and 1.8).

SURFACE-PHASE LEARNING INTENTIONS FOR EACH COMPONENT OF MATHEMATICS LEARNING

Learning Intentions	Conceptual Understanding	Procedural Knowledge	Application of Concepts and Thinking Skills
Unistructural (one idea)	I am learning that the purpose of a graph is to represent data gathered to answer a question.	I am learning that there are ways to represent data using graphs.	I am learning that I can use data to answer questions that I want to investigate.
Multistructural (many ideas)	I am learning that different questions produce different types of data.	I am learning that there are multiple ways to represent data using graphs.	I am learning that there are specific characteristics of my graph that represent my data.

Figure 1.7

SURFACE-PHASE SUCCESS CRITERIA FOR EACH COMPONENT OF MATHEMATICS LEARNING

Success Criteria	Conceptual Understanding	Procedural Knowledge	Application of Concepts and Thinking Skills
Unistructural (one idea)	I can describe how a graph represents data.	I can describe the parts of a graph.	I can create a question that generates data.
Multistructural (many ideas)	I can identify specific questions and data that are represented by different types of graphs.	I can give examples of different types of graphs.	I can list the characteristics of a graph that would answer my question.

Figure 1.8

Deep Learning in the Intermediate Mathematics Classroom

Biggs and Collis (1982) conceptualize deep learning as identifying relationships between concepts or ideas. Learners at the deep level of the learning process focus on relationships and relational thinking about concepts, procedures, and applications. Returning to the perimeter and area of a rectangle problem mentioned previously, learners are able to compare and contrast the procedure for finding the area and perimeter in different contexts. Conceptually, learners deepen their understanding of the length, width, and sides of a rectangle and the relationship of these values to the area and perimeter. In other words, learners see that the relationship between the sides is more important than whether the side is called a length or width in a rectangle. Likewise, they recognize the equivalent dimensions in a square. They can analyze a specific situation and determine the best approach to finding the perimeter or area without specific guidance on which approach is most efficient and effective. The development of relational thinking paves the way for transferring these concepts and thinking, or as Biggs and Collis (1982) call it, *extending thinking*. The learning intentions and success criteria should reflect this level of thinking or readiness for our learners (Figures 1.9 and 1.10).

DEEP-PHASE LEARNING INTENTIONS FOR EACH COMPONENT OF MATHEMATICS LEARNING

Learning Intentions	Conceptual Understanding	Procedural Knowledge	Application of Concepts and Thinking Skills
Relational (related ideas)	I am learning that the specific context of the situation determines how to best represent the relationship (e.g., the type of graph).	I am learning the relationship between the data generated by my question and the type of graph.	I am learning that how I ask a question influences the data I will need to collect.

Figure 1.9

DEEP-PHASE SUCCESS CRITERIA FOR EACH COMPONENT OF MATHEMATICS LEARNING

Success Criteria	Conceptual Understanding	Procedural Knowledge	Application of Concepts and Thinking Skills
Relational (related ideas)	I can explain the relationship(s) between the context of the question and the type of graph.	I can justify the type of graph I created from a given set of data.	I can use a graph to answer questions.

Figure 1.10

Transfer Learning in the Intermediate Mathematics Classroom

The next step in the SOLO progression is for the learner to transfer learning to different contexts. At the extended level of thinking, learners formulate big ideas and generalize their learning to a new domain. For example, an extended abstract thinker might predict how to find the perimeter, and even the area, of an irregular polygon. Learners at this level may begin to generalize this to other two-dimensional geometric shapes, recognizing that there are dimensions that maximize the area of a specific shape. Learners will begin to extend their thinking by using procedures in very different situations.

Being clear about the learning intentions and success criteria is just as important in extending student ideas as with the previous levels of thinking (Figures 1.11 and 1.12).

TRANSFER-PHASE LEARNING INTENTIONS FOR EACH COMPONENT OF MATHEMATICS LEARNING

Learning Intentions	Conceptual Understanding	Procedural Knowledge	Application of Concepts and Thinking Skills
Extended abstract (extending ideas)	I am learning that graphs represent numeric relationships in authentic situations.	I am learning how graphs can be interpreted and represent characteristics of the data.	I am learning how I can use graphs to make predictions.

Figure 1.11

TRANSFER-PHASE SUCCESS CRITERIA FOR EACH COMPONENT OF MATHEMATICS LEARNING

Success Criteria	Conceptual Understanding	Procedural Knowledge	Application of Concepts and Thinking Skills
Extended abstract (extending ideas)	I can interpret graphs to develop characteristics of the data I collected.	I can rearrange my own data to focus on a specific question.	I can explain a prediction based on the analysis and interpretation of a graph.

Figure 1.12

With clear learning intentions and success criteria in place, we must design learning experiences and challenging mathematics tasks that result in students engaging in both mathematical content and processes at the right level of thinking. This brings us to the question of rigor.

Differentiating Tasks for Complexity and Difficulty

As we have noted, there are three phases to student learning: surface, deep, and transfer. Teachers have to plan tasks that provide students opportunities to learn and progress through these stages, as well as the flexibility to return to different phases of the learning when necessary. When students experience a "Goldilocks" challenge, the effect size is 0.74. A Goldilocks challenge is not too hard and not too boring. For example, if learners need additional surface learning around some aspect of procedural knowledge or conceptual understanding, we have the flexibility to go back, provide that instructional support, and then continue in the learning. The type of task matters as students move along in their thinking from surface to deep to transfer. In *Visible Learning for Mathematics*, we shared the Common Core State Standards for Mathematics definition of rigor as the balance of conceptual learning, procedural skills and fluency, and application. This is a good definition when applied to mathematics instruction, curricula, and learning as a whole. But we also want to address the appropriate challenge of any individual mathematical *task*. In this book, we are using the term **rigor** to mean the balance of complexity and difficulty of a mathematical task.

As soon as someone mentions "rigorous tasks," we mentally formulate what those are in our own classrooms. Is rigor completing 50 problems for homework? Is rigor engaging in a mathematics brainteaser? To effectively design rigorous mathematics tasks that align with our learning intentions and success criteria, we have to better understand what is meant by difficulty and complexity. *Difficulty* is the amount of time, effort, or work expected of the student, whereas *complexity* is the level of thinking, the number of steps, or the abstractness of the task. We can differentiate by adjusting the level of difficulty and/or complexity for any task regardless of whether the task focuses on conceptual understanding, procedural knowledge, or application. In Ms. Showker's classroom, learners are expected to construct a line graph and a bar graph

---

With clear learning intentions and success criteria in place, we must design learning experiences and challenging mathematics tasks that result in students engaging in both mathematical content and processes at the right level of thinking.

EFFECT SIZE FOR "RIGHT" LEVEL OF CHALLENGE = 0.74

Rigor is the level or balance of difficulty and complexity of any given mathematical task.

from data. For this specific success criterion, Ms. Showker adjusted the difficulty of the task, while maintaining the level of complexity set by the success criteria, by providing learners with different levels of scaffolding as follows.

Adjustments to the Task Difficulty

☐ Graph paper with rows or columns drawn for each category, and options for values of pictures

☐ Stamps or stickers to use in constructing the graphs

☐ Graph paper with the axes drawn and scaled (learners need to graph the data, label the axes, provide a key if necessary, and add a title)

☐ Graph paper with the axes drawn, but not scaled

☐ Graph paper with a checklist of the components needed for each graph

☐ A blank sheet of paper

All learners in the classroom worked toward the same success criteria and level of complexity. What differed was the level of difficulty to ensure that all learners had access to the learning. As learners develop greater procedural knowledge and conceptual understanding, the level of difficulty can be increased by gradually removing the scaffolding.

We do not believe that teachers can radically impact student learning by making them do a lot more work. Practicing hundreds of division problems (increased difficulty) will not extend their thinking. Similarly, asking students to engage in a task that is far too complex or not complex enough for their current level of thinking can also reduce the impact on student learning. Instead, we should balance difficulty and complexity in the design of learning tasks. Throughout this book, we will return to the concepts of difficulty and complexity as we discuss the various strategies and tasks our three profiled teachers use and share how they can adjust the difficulty and complexity of those tasks to meet the needs of all learners.

Approaches to Mathematics Instruction

Just as task design is an important consideration in the Visible Learning classroom, learners need to experience a *wide range* of tasks if they are going to become assessment-capable visible mathematics learners. They need opportunities to work with their teacher, with their peers, and independently so that they develop the social and academic skills necessary to continue to learn on their own. Although Ms. Showker decided to use a student-led dialogic approach, this is just one of four approaches to mathematics instruction. Three additional approaches are direct/deliberate instruction, teacher-led dialogic instruction, and independent learning.

Direct/Deliberate Instruction. Direct/deliberate instruction, commonly referred to as direct instruction, has a negative reputation in education. This approach is mistakenly assumed to be synonymous with lecture or "telling" students information. That is not the case. Direct/deliberate instruction involves the following:

- Activation of prior knowledge

- Introduction of the new concept or skill

- Guided practice of the concept or skill

- Feedback on the guided practice

- Independent practice

To limit one's understanding of direct/deliberate instruction to highly scripted programs or lecture is to overlook the practices that make it highly effective for developing surface-level knowledge. With an effect size of 0.60, direct/deliberate instruction offers a pedagogical pathway that provides students with the modeling, scaffolding, and practice they require when learning new concepts and skills, as further explained by Hattie (2009):

> When we learn something new . . . we need more skill
> development and content; as we progress, we need more

EFFECT SIZE FOR DIRECT/ DELIBERATE INSTRUCTION = 0.60

Guided practice involves the teacher and the students collaboratively engaged in problem solving. This helps the teacher and learners determine when students are ready to work independently.

EFFECT SIZE FOR
SCAFFOLDING =
0.82

EFFECT SIZE
FOR DELIBERATE
PRACTICE = 0.79

EFFECT SIZE FOR
FEEDBACK = 0.70

Teaching Takeaway

These approaches are in no particular order. Using the right approach, at the right time increases our impact on student learning in the mathematics classroom.

EFFECT SIZE FOR
QUESTIONING
= 0.48

EFFECT SIZE
FOR SELF-
VERBALIZATION
AND SELF-
QUESTIONING
= 0.55

connections, relationships, and schemas to organize these skills and content; we then need more regulation or self-control over how we continue to learn the content and ideas. (p. 84)

Teacher-Led Dialogic. As learners develop the skills to engage in deepening dialogue, teacher-led dialogic instruction allows the teacher to be present in student discussions about mathematics, facilitating the process to scaffold student conversation. In the end, the teacher will fade his or her support as students develop the necessary skills to take over and lead the conversations on their own. Teacher-led dialogic instruction does not require direct/deliberate instruction first. Instead, this approach requires learners to possess the surface knowledge necessary to engage in deeper dialogue. For example, a teacher may utilize a teacher-led dialogic approach as she introduces the reasoning necessary to recognize congruent angles and triangles. Over time, after modeling the type of questioning and reasoning, the teacher's role in this dialogue will lessen, gradually releasing the students to more independent work (i.e., less dependent on the teacher).

Student-Led Dialogic. Students have a way of making themselves understood by their peers. In other words, students' thoughts and explanations can propel the learning of their peers. Whether solving problems, providing feedback, or engaging in reciprocal teaching, the collaborative act of peer-assisted learning in mathematics benefits all students in the exchange. In student-led dialogic learning, the role of the teacher is to organize and facilitate, but it is the students who are the ones that lead the discussion.

Independent. The learning continues, and in fact deepens, when students are able to employ what they have been learning. This can occur in three possible ways (Fisher & Frey, 2008):

- Fluency building
- Spiral review
- Extension

Fluency building is especially effective when students are in the surface learning phase and need spaced practice opportunities to strengthen automaticity. For instance, students who play online mathematics games, or engage in mathematics problem solving independently, are engaged in fluency-building independent learning.

Spiral review is one in which students revisit previous content for which they need additional support or to activate prior knowledge for the day's learning intention and success criteria.

Extension promotes transfer and occurs when learners are asked to use what they have learned in a new way. Independent learning through extension includes writing about mathematics, teaching information to peers, and engaging in mathematics investigations.

Checks for Understanding

Checks for understanding offer both teachers and learners the opportunity to monitor the learning process as students engage in challenging tasks and progress toward the learning intention. To ensure the learning is visible in our mathematics classroom, we must have the necessary information about student progress so that we provide effective feedback. In addition, learners must also have the necessary information about their progress so that they can effectively monitor progress and adjust their learning. Using the success criteria as a guide, checks for understanding include any strategies, activities, or tasks that make student thinking visible and allow both the teacher and learner to observe learning progress. When we are planning, developing, and implementing checks for understanding, two essential questions should guide our thinking:

Guiding Questions for Creating Opportunities to Respond:

1. What checks for understanding will tell me and my learners how they are progressing in their learning related to the learning intention(s) and success criteria?

2. What are we going to do with this information that will help students with their next steps in learning this content?

Checks for understanding give us information about the impact of our teaching and should be driven by the learning intention and success criteria for that particular lesson or learning experience. In other words,

> EFFECT SIZE FOR
> HELP SEEKING
> = 0.72

> EFFECT SIZE FOR
> SELF-REGULATION
> STRATEGIES = 0.52

There is no one way to teach mathematics. We should not hold any influence, instructional strategy, action, or approach to teaching and learning in higher esteem than students' learning.

Teaching Takeaway

Unless we, as teachers, have clear success criteria, we are hardly likely to develop good checks for understanding for our learners.

checks for understanding are used as feedback for teachers about their teaching. For example, if the success criteria say to *describe*, the check for understanding should focus on or provide deliberate practice in *describing*. Someone teaching mathematics in the Visible Learning classroom should focus on assessment for the purpose of informing instructional decisions and providing feedback to learners. The following assumptions inform our collective understanding about teaching and learning:

1. Assessment occurs throughout the academic year, and the results are used to inform the teacher and the learner. Each period, time is set aside to understand students' mathematics learning progress and provide feedback to learners.

2. A meaningful amount of time is dedicated to developing mathematics content and processes. Across every unit, students engage in sustained, organized, and comprehensive experiences with all of the components: conceptual understanding, procedural knowledge, and application of concepts and thinking skills.

3. Solving problems and discussing tasks occurs every class period. These events occur with the teacher, with peers, and independently.

Profiles of Three Teachers

In addition to the videos accompanying each chapter of this book, we will follow the practices of three teachers throughout the remaining chapters. Just as we have provided specific examples throughout this chapter and in the videos, we will devote more time to take an in-depth look into the classrooms of these three elementary mathematics teachers. We will give you a front-row seat as they make specific, intentional, and purposeful decisions in teaching mathematics in the Visible Learning classroom.

Beth Buchholz

Beth Buchholz is a third grade teacher in Indiana. Although she has taught second and fifth grade, Ms. Buchholz finds that third graders are her niche. Her experience across multiple grade levels has allowed her to understand where her learners are coming from in their mathematics learning journey and where they are headed in fifth grade: "I believe that I have an advantage by seeing the vertical perspective on mathematics knowledge, procedures, and concepts, as well as the learning progressions of my students." Ms. Buchholz has spent her entire 15-year teaching career in this urban, very diverse school district. She teaches English language arts, mathematics, science, and social studies in a room with 33 learners. Once a week, her grade-level team gets an extended planning time of 94 minutes. The other four days allow for 45 minutes of planning. Approximately 22% of her learners speak a language in addition to English. Her school qualifies as a school-wide Title I school, with 73% of the students qualified for free lunch (a measure of poverty). In addition, 16% of the students receive special education services. Ms. Buchholz is working to incorporate her own learning into her teaching and her students' learning. She is working toward her Math Specialist certificate at a local university. As she notes, "There is so much to learn about teaching and learning in mathematics that I sometimes get overwhelmed. I have to keep my focus on what works best and what evidence I have that a particular approach works best."

Hollins Mills

Hollins Mills teaches fourth grade in Virginia. She is in her third year of teaching in this small district just a few miles from a major state university. Her first 2 years were not as smooth as she had envisioned during her teacher preparation program. Upon entering a school district in which 46 different languages are spoken, Ms. Mills had to make quick adjustments to her classroom to ensure that all learners had access to mathematics concepts and thinking. In Virginia, learners must pass an end-of-course assessment that is also factored into the teacher evaluation system. However, Ms. Mills does not let this distract from planning, developing, and implementing challenging mathematics tasks. She says, "I believe that if we focus on learning, the test scores take care of themselves. This may be naive, but it seems to be working." For Ms. Mills, mathematical process skills are essential to learning and she believes that all students can master this type of thinking. She works very hard to emphasize mathematical thinking each and every day, pushing back against a focus only on procedures (e.g., memorizing algorithms or mnemonics). Her strengths are in teaching mathematics and science. She teaches mathematics, as well as science, to all of the fourth graders at her school. This allows her to stretch her learners' thinking by applying concepts and thinking skills to science. There are about 24 students in each fourth grade classroom. Her learners are quite diverse (42% are white, 33% are black, and 12% are Hispanic), and 15% speak Spanish as their heritage language. Approximately 60% of the students in the school qualify for free lunch, and 13% of the students receive special education services.

Katy Campbell

Katy Campbell is a fifth grade teacher in rural California. She is part of a five-person grade-level team that meets regularly to discuss student progress in learning. The team also includes a special education teacher who provides accommodations and modifications and small group instruction for the 119 fifth graders at her school. This is a large elementary school in which each grade level has at least five sections. The 24 learners in Ms. Campbell's classroom are diverse, with 83% of them qualifying for free or reduced school lunch. Fifty-five percent of her students are female, and 45% are African American, 35% are Latino/Hispanic, 15% are Asian Pacific Islander, 4% are Caucasian, and 1% identify as other. Of her 24 students, 6 are students with disabilities. Overall, her learners have a wide range of instructional needs. Ms. Campbell has enrolled in an online course on differentiated instruction to enhance her skills to meet each student's instructional needs. She says, "I want to make sure I do not leave any stone unturned in terms of ideas or ways to reach my learners." Ms. Campbell is entering her tenth year as a teacher, having only taught fifth grade. Although she is very comfortable with the Common Core State Standards, she pays very close attention to her impact on student learning. She says, "In addition to meeting with my grade-level team, I spend a lot of time looking at student work. I want to know exactly what they get and don't get so that I can use the instructional time as effectively as possible. Plus, if I cannot connect with them, I have to find another way."

These three teachers, although in different regions and contexts, operate under three important assumptions:

1. There is no one way to teach mathematics or one best instructional strategy that works in all situations for all students, but

there is compelling evidence for tools that can help students reach their learning goals.

2. We should not hold any influence, instructional strategy, action, or approach to teaching and learning in higher esteem than students' learning.

3. Effective teaching and learning requires establishing clear learning intentions and success criteria, designing learning experiences and challenging mathematics tasks, monitoring student progress, providing feedback, and adjusting lessons based on the learning of students.

In the chapters that follow, you will encounter these three teachers and view the lesson plans they have developed for themselves. In order to establish a predictable pattern for displaying this information, we will use the Planning for Clarity questions (see Figure I.4). Lessons based on these guiding questions are not meant to be delivered in a strictly linear fashion; rather, they are intended to serve as a way to guide your thinking about the elements of the lesson. In addition, through the videos accompanying this book, you will more briefly meet a number of teachers from other grade levels whose practices illustrate the approaches under discussion. Although no book on lesson planning could ever entirely capture every context or circumstance you encounter, we hope that the net effect is that we provide a process for representing methods for incorporating Visible Learning for mathematics consistently in your classroom.

Reflection

Mathematics instruction that capitalizes on Visible Learning is established upon principles of learning. Recognizing that learners develop procedural knowledge, improve conceptual understanding, and apply concepts and thinking by engaging in surface, deep, and transfer learning allows us to intentionally and purposefully foster increasingly deeper and more sophisticated types of thinking in mathematics. This focus on the individual learner makes this approach inclusive of all learners, including those with language or additional learning needs. Teaching mathematics in the Visible Learning classroom means leveraging high-impact instruction to accelerate student learning through surface, deep, and transfer phases of learning by engaging them in strategies, actions, and approaches to learning at the right time and for the right content. These challenging learning tasks have clear learning intentions and success criteria that allow students to engage in those tasks in a variety of ways and with a variety of materials. Learning becomes visible for the teacher and the students. In other words, an assessment-capable visible mathematics learner notices when he or she is learning and is proactive in making sure that learning is obvious. As we engage in discussions about mathematics learning in this book, we will return to these indicators that students are visible mathematics learners to explore how they might look in the classroom.

1. Take a moment and develop your own explanation of teacher clarity. What does teacher clarity look like in your mathematics classroom?

2. Using an upcoming lesson plan as an example, what components of mathematics instruction are you focusing on in the lesson? How does your lesson incorporate all or some of the following?

   a. Making sense of problems and persevering in solving them

   b. Reasoning abstractly and quantitatively

   c. Constructing viable arguments and critiquing the reasoning of others

   d. Modeling with mathematics

   e. Using appropriate tools strategically

   f. Attending to precision

   g. Looking for and making use of structure

   h. Looking for and expressing regularity in repeated reasoning

3. Using that same lesson plan, how will you or could you adjust the difficulty and/or complexity of the mathematics tasks to meet the needs of all learners?

4. Give some examples of learners engaged in surface learning, deep learning, and transfer learning. What are the observed learning outcomes of these students? What learning experiences best support learners at each level?

TEACHING FOR THE APPLICATION OF CONCEPTS AND THINKING SKILLS 2

CHAPTER 2 SUCCESS CRITERIA:

(1) I can describe what teaching for the application of concepts and thinking skills in the mathematics classroom looks like.

(2) I can apply the Teaching for Clarity Planning Guide to teaching for application.

(3) I can compare and contrast different approaches to teaching for application.

(4) I can give examples of how to differentiate the complexity and difficulty of mathematics tasks designed for application.

Assessment-capable visible learners in the mathematics classroom use mathematics in situations that require the application of mathematics concepts and thinking skills. How efficiently and effectively this occurs depends on the learners' conceptual understanding and procedural knowledge. When planning for clarity (see Figure I.4), we begin with the end in mind and we ask ourselves, "What do I want my students to learn?"

In this chapter, we take the same approach. Ms. Buchholz, Ms. Mills, and Ms. Campbell focus on the end goal for each of their learners as they design meaningful learning experiences. All three teachers expect their learners to apply mathematics concepts and thinking skills to authentic situations. Thus, our journey begins with how these three teachers, by design, teach for this purpose. The QR codes in the margin provide video examples of application in action from other mathematics classrooms. In Chapters 3 and 4, we will go back in time and look at how these classrooms got here.

The nature of the application of concepts and thinking skills differs *across* the three classrooms and *within* the three classrooms. How these teachers approach this purpose depends on the learning needs of the students in their classroom. Therefore, you will see that Ms. Buchholz, Ms. Mills, and Ms. Campbell adjust the rigor—or complexity and difficulty—of the application task, depending on where their learners currently are in the learning process (e.g., surface, deep, or transfer). For example, Ms. Buchholz adjusts the rigor of her application task for learners who need additional surface learning around the specific application task. Likewise, Ms. Mills and Ms. Campbell adjust the rigor of their application task to support learners who have gaps in their conceptual understanding. In all three classrooms, learners apply concepts and thinking skills to authentic scenarios. As we journey through these three classrooms, pay special attention to how each teacher differentiates the complexity and difficulty of the mathematics tasks so that all learners have access and the opportunity to apply concepts and thinking skills.

Ms. Buchholz and the Relationship Between Multiplication and Division

Ms. Buchholz's students are ready to apply their conceptual understanding and procedural knowledge about the operation of multiplication to a new context. The class has worked with single digits to answer the following

Teaching Takeaway

Teaching mathematics in the Visible Learning classroom uses the right approach at the right time.

Teaching Takeaway

We can differentiate mathematics tasks by adjusting the complexity and difficulty of the task.

unit essential questions: What is multiplication? What is division? How are they related? and How do we engage in the work of mathematicians? Along the way, they have posed their own questions to which they have pursued answers: Will multiplication ever result in a smaller product? How can you multiply to solve a division problem? Can you decompose the dividend or the divisor to make division more efficient? How do you compensate if you adjust the factors in a multiplication problem?

EFFECT SIZE FOR PRIOR ABILITY = 0.94

Now, Ms. Buchholz is presenting her class with transfer essential questions:

- How can we use what we know about multiplying and dividing small numbers to multiply and divide large numbers?

EFFECT SIZE FOR QUESTIONING = 0.48

- How can we apply the properties (associative, commutative, and distributive) to problems with large numbers?

They will explore these questions through the Auditorium task.

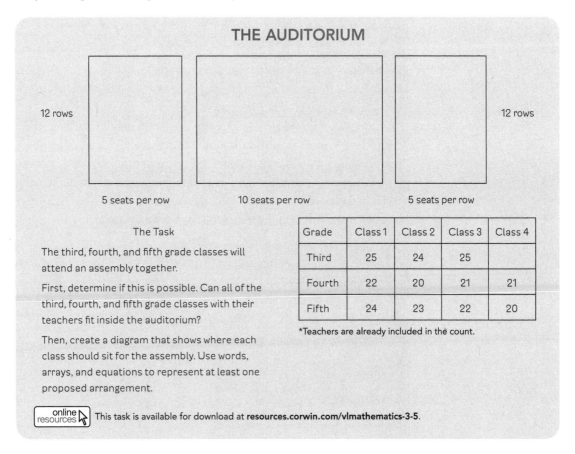

THE AUDITORIUM

12 rows   12 rows

5 seats per row   10 seats per row   5 seats per row

The Task

The third, fourth, and fifth grade classes will attend an assembly together.

First, determine if this is possible. Can all of the third, fourth, and fifth grade classes with their teachers fit inside the auditorium?

Then, create a diagram that shows where each class should sit for the assembly. Use words, arrays, and equations to represent at least one proposed arrangement.

Grade	Class 1	Class 2	Class 3	Class 4
Third	25	24	25	
Fourth	22	20	21	21
Fifth	24	23	22	20

*Teachers are already included in the count.

online resources ⮡ This task is available for download at **resources.corwin.com/vlmathematics-3-5**.

EFFECT SIZE FOR
PERCEIVED TASK
VALUE = 0.46

A challenging and fascinating quality of mathematics is the ever-expanding number system. Ms. Buchholz knows this quality can make some students (and teachers) feel overwhelmed, as though the rules are always changing and the quantity of things to learn in math is always growing. She embraces this quality of math by creating explicit opportunities for her students to transfer what they already know to new situations, new operations, new sizes of numbers, and new types of numbers. She also makes sure her instruction centers on vertical mathematical truths and big ideas rather than narrow facts that can lead to misconceptions. In forming the transfer essential questions, Ms. Buchholz intentionally creates an opportunity for students to return to concepts and skills that remained at the surface or deep level earlier in the unit. The application task engages students in exploring multidigit multiplication and division while also providing needed additional yet different experiences to reach mastery with single digits.

What Ms. Buchholz Wants Her Students to Learn

To create her unit, Ms. Buchholz studies each of the standards related to multiplication and division and chunks them so that each is a building block to the next. Overall, Ms. Buchholz begins with one-digit numbers and single-step, single-operation contextualized problems and then moves on to multidigit, multistep, multioperation contextualized problems. This sequence mirrors the SOLO Taxonomy model, beginning with one or a few aspects of the concept (unistructural), then several aspects (multistructural), then integrating all aspects into a meaningful whole (relational), and finally, generalizing and transferring that whole to new contexts (extended abstract) (Biggs & Collis, 1982).

EFFECT SIZE FOR
COGNITIVE TASK
ANALYSIS = 1.29

Ms. Buchholz's first application lesson will focus on the standards that make the integration and transfer of multiplication concepts and skills visible to her students.

INDIANA ACADEMIC STANDARDS

3.C.5. Multiply and divide within 100 using strategies, such as the relationship between multiplication and division (e.g., knowing that $8 \times 5 = 40$, one knows $40 \div 5 = 8$), or properties of operations.

3.AT.3. Solve two-step real-world problems using the four operations of addition, subtraction, multiplication, and division (e.g., by using drawings and equations with a symbol for the unknown number to represent the problem).

Ms. Buchholz is helping her learners develop the following Standards for Mathematical Practice:

- Construct viable arguments and critique the reasoning of others.

- Look for and make use of structure.

Learning Intentions and Success Criteria

Ms. Buchholz uses learning intentions and success criteria each day to help her students practice self-monitoring, self-evaluation, and goal setting. When students evaluate their progress toward the learning intentions, find evidence of the success criteria in their work, and make plans for next steps, their learning becomes visible to themselves and others. The learning intentions integrate the transfer essential questions and standards to tell students what and why they are learning today. Ms. Buchholz's approach is to develop learning intentions for content, language, and social dimensions of this application experience. Dividing learning intentions into *content*, *language*, and *social* varieties can provide teachers and students alike a clearer sense of the day's expectations. **Content learning intentions** answer the question "What is the math I am supposed to use and learn today?" **Language learning intentions** give teachers a space to lay out the language demands of the day: Are students developing new academic or content vocabulary, are they practicing recently developed vocabulary within proper linguistic structures, or are they utilizing those structures toward their actual communicative functions? This is not limited to verbal communication and can include written or visual representations of mathematical thinking. **Social learning intentions** allow teachers to develop and leverage social and sociomathematical norms within their classroom culture.

> EFFECT SIZE FOR TEACHER CLARITY = 0.75

Content learning intentions: What is the math I am supposed to use and learn today?

Language learning intentions: How should I communicate my mathematical thinking today?

Social learning intentions: How should I interact with my learning community today?

Ms. Buchholz's learning intentions for this lesson are as follows:

Content Learning Intention: I am learning how strategies for multiplying and dividing small numbers can be transferred and revised to multiply and divide large numbers.

Language Learning Intention: I am learning how mathematical properties can be used to describe and defend multiplication and division strategies.

Social Learning Intention: I am learning to appreciate the contributions of each learner and the connections among others' reasoning and my own.

The success criteria break down what it will look like and sound like when students have mastered the learning intentions. Ms. Buchholz's success criteria for the day are as follows:

- ☐ I can visually represent partial products in an array.
- ☐ I can decompose and compose products in multidigit multiplication using partial products.
- ☐ I can apply the distributive, commutative, and associative properties to multiplication of large numbers.
- ☐ I can transfer and revise efficient strategies for multiplication and division of large numbers.

Ms. Buchholz introduces the learning intentions and success criteria after activating students' prior knowledge about partial products, since this phrase is significant to making sense of the success criteria. She will return to the learning intentions and success criteria as students share their many strategies and solutions and again as students begin to assemble their math portfolios.

Activating Prior Knowledge

One equation is written in the corner of the whiteboard: $8 \times 5 = 40$. The expression 8×3 is written in the middle of the board. The room is silent as

---

Teaching Takeaway

Success criteria should articulate the evidence learners must show to demonstrate if they have met the learning intentions or their growth toward the learning intentions.

EFFECT SIZE FOR STRATEGY TO INTEGRATE WITH PRIOR KNOWLEDGE = 0.93

students raise their thumbs. Ms. Buchholz gathers solutions to 8×3 and records them on the board: 24 and 25. She asks, "Who will defend 24?"

Naum begins, "I know 8×2 is 16 and one more set of 8 is 24." Ms. Buchholz records the following:

$$8 \times 2 = 16$$
$$8 \times 1 = 8$$
$$16 + 8 = 24$$

"Who has another way to confirm Naum's solution?" Ms. Buchholz asks. The students know what to expect in this routine, called a number talk (Humphreys & Parker, 2015; Parrish, 2014). They know they are practicing mental math strategies because this form of procedural fluency is one of their year-long goals.

"I used the problem we just did: $8 \times 5 = 40$," says Mario. "But that's five groups of 8 and we only need three. So I took away two groups of 8 or 16." Ms. Buchholz records his answer:

$$8 \times 5 = 40$$
$$8 \times 2 = 16$$
$$40 - 16 =$$

Mario watches and pauses. He says, "Well, really I thought 16 and what makes 40 because I prefer to think addition when I subtract. I can make 10 easily. So that's 24: $8 \times 3 = 24$." Ms. Buchholz erases "$40 - 16 =$" and writes "$16 + \rule{1cm}{0.4pt} = 40$" and then fills in 24 in the blank. She uses this opportunity to model ways to record mental math strategies.

The teacher says, "When I read this equation, I can think like Mario did: 'The difference between five groups of 8 and two groups of 8 is three groups of 8, or 24.'" Ms. Buchholz rephrases Mario's strategy to highlight the connection between subtraction and the solution. She knows using meaningful language helps all of her students make sense of mathematical notation.

"I'd like to revise my thinking, Ms. Buchholz," Akeiyla adds. "I thought about it like Naum but I added 16 and 8 wrong. I agree the solution

Teaching Takeaway

Building learners domain-specific or tier vocabulary is necessary for a successful number talk or mathematics discussion.

EFFECT SIZE FOR SELF-REGULATION STRATEGIES = 0.52

is 24." Ms. Buchholz crosses off 25, and then she writes "8 × 3 = 24" beneath 8 × 5 = 45 in the corner and erases the rest of the work.

Ms. Buchholz continues, "Mario said he used the previous problem to help him. Think about how you can use the two previous problems to help you fluently solve 8 × 8." She records the final problem in the number string. She allows wait time until the majority of students have raised their thumbs. Some students are showing a thumb and an index finger to indicate they have solved the problem in two different ways.

EFFECT SIZE FOR
EVALUATION AND
REFLECTION = 0.75

After students share solutions and defend them, the class concludes that 8 × 8 = 64. Ms. Buchholz rewrites two of the shared strategies in a new format:

$$8 \times 8 = 8 \times (5 + 3) = (8 \times 5) + (8 \times 3)$$
$$8 \times 3 = 8 \times (2 + 1) = (8 \times 2) + (8 \times 1)$$

She says, "Here is Briera's strategy for 8 × 8 and Naum's strategy for 8 × 3. I have rewritten them horizontally with parentheses to show the ways they decomposed one factor. What do you notice?" Ms. Buchholz pauses for students to study and think about the equations. "Talk with your elbow partner about your noticings," she says. The students talk and point. After 1 minute, Ms. Buchholz begins the whole-group share.

"I noticed that both Briera and Naum decomposed one factor, but they still multiplied each of the decomposed numbers by the other factor," Ty shares.

"I noticed that too. And then after multiplying by the other factor, both of them added the new products to get one big product," Pakmo adds.

She facilitates making connections to the academic math language: "We've used this strategy before in our number talks and work. Each of those smaller products is called a partial product because it's part of the big product you're trying to solve. What property does this strategy rely on?" She writes the important vocabulary on the board.

"Distributive property," says Sabina.

Ms. Buchholz writes this next to the equations and continues.

> You've used the distributive property with many small
> numbers for multiplication. Today, we're going to explore
> what happens with the distributive property and other
> multiplication strategies when we work with larger numbers.
> We're going to compare strategies and solutions in order to
> make connections about how to transfer strategies to multiply
> larger numbers.

She restates the learning intentions as part of her transition from activating prior knowledge to introducing the tasks.

Ms. Buchholz posts an enlarged copy of the Auditorium task. She instructs the students to read the task and think about what information they need to solve the problem. She is using the instructional strategy called **Bansho** (Curriculum Services Canada, 2011; Kuehnert, Eddy, Miller, Pratt, & Senawongsa, 2018) in order to help students meet the learning intentions. Bansho, or board writing, emphasizes the comparison of strategies to synthesize big ideas. The Auditorium task is an open question (Small, 2012), which means it is open middled (many strategies) and open ended (many solutions). With so many possible strategies and solutions, Ms. Buchholz wants to target the synthesis of big ideas, which will answer the transfer essential questions.

After she allows some think time, Ms. Buchholz records students' noticings and important vocabulary (such as *array*, *seats per row*, and *section*). These notes, along with the recordings from the number talk, will remain on the board as a reference. The class discussion lets Ms. Buchholz hear that her students understand the task and are ready to problem solve. She says, "We will use four success criteria to evaluate our work today." She reads the four success criteria and adds the following instructions:

> As you work, keep these success criteria in mind. At the end
> of our work time, you will evaluate your process and product
> using these success criteria and you will identify evidence to
> prove you've met each.

Today, you will work with a team. There are many materials available for problem solving, including graph paper, colored pencils, array cards, tiles, calculators, your multiples number charts, number lines, and your mathematical toolboxes.

Ms. Buchholz has intentionally chosen the groups in advance for today. She also wants to give students an opportunity to practice making decisions about what materials to use in their mathematics learning. She has plans for holding students accountable for today's learning when they select evidence to support their self-evaluation.

Scaffolding, Extending, and Assessing Student Thinking

As students spread out to work, Ms. Buchholz refers to her observation checklist, which includes each student's name, a list of anticipated strategies (Smith & Stein, 2011), the success criteria, and her planned questions. In her conferences with students, Ms. Buchholz asks the following questions to scaffold, extend, and assess student thinking:

Teaching Takeaway

When learners work collaboratively during a mathematics task, personal accountability in those tasks can be used to monitor students' progress toward the learning intentions and success criteria.

- How could you represent the size of each class using arrays and equations?

- How does rotating this array affect your solution?

- Why are these small arrays equal to this one large array? How could you represent this equality?

- How are you using partial products?

- How are you using other multiplication strategies from our anchor chart (making friendly numbers, skip counting, repeated addition)?

- Where do you see the associative property? Commutative property? Distributive property?

- How are you thinking about both multiplication and division in this problem?

Teaching Takeaway

Teaching mathematics in the Visible Learning classroom requires that we, as teachers, gather evidence to measure our impact on student learning.

Based on her conferences and observations, Ms. Buchholz selects and sequences five groups of students to share:

1. Sabina and Machele used tiles to represent each person and tile colors to represent each class. To record, they drew lines to represent the seating area for each class and wrote equations with colored numbers to match each class.

2. Adoni, Adam, and Mario cut arrays out of graph paper and glued them to create the auditorium sections. They recorded equations for each array and for the overall sections.

3. Kayvion and Frances drew the area of each array on the diagram and labeled each class. They created an equation to show the total number of seats in the auditorium based on the number of full seats and empty seats.

4. Lobsang and Leo were the only pair to split the middle section in half to create two additional 5×12 arrays. Their equations show small arrays creating four large arrays, rather than three large arrays.

5. Naum, Pakmo, and Grace sketched arrays on the diagram and colored extra seats a different color. Their equations include these extra seats as addends.

Teaching for Clarity at the Close

In the consolidation phase of Bansho, Ms. Buchholz displays the student strategies she selected in sequence from left to right on the board, next to the notes from activating prior knowledge. The strategy furthest left relies on concrete representations (tiles), and more abstract strategies or complex ideas are posted to the right. Each group of students presents their strategy, representations, and reasoning. Students in the audience ask clarifying questions. Ms. Buchholz annotates next to each question with important math vocabulary and concise explanations such as "Rotating the array is the commutative property," "Small arrays are represented inside parentheses," and "The total area of the large array equals the sum of the area of the small arrays."

After all five groups share, Ms. Buchholz facilitates the class discussion in order to highlight key ideas, strategies, models, and properties as well as to begin to answer the transfer essential questions. She records these to the right of the student work.

Teaching Takeaway

Make student thinking visible. This allows us to see learning through the eyes of our students and students see themselves as their own teachers.

Teaching Takeaway

At the end of an instructional block, we must provide closure to the teaching and learning time. This promotes the consolidation of the learning.

EFFECT SIZE FOR ELABORATIVE INTERROGATION = 0.42

"What do you notice is the same across all of the strategies and representations?" Ms. Buchholz begins.

"Everyone decomposed the large arrays into small arrays because the numbers were friendlier. Even Adoni's team thought about the large 12×5 array as a 10×5 array and then a 2×5 array, even though it meant cutting up more paper," Dillon notes.

"Everyone used the distributive property too. You can see it with the arrays but you can also see it in some of the equations, like Leo and Lobsang's," adds Grace.

Ms. Buchholz jumps at this opportunity to highlight an important feature of the distributive property: All small arrays must share one side with the same length or width. She points this out when Grace states the misconception that you can break a large array into any size arrays to use the distributive property. She says, "Let's look at Sabina and Machele's arrays versus Adoni, Adam, and Mario's arrays. What do you notice is different about them?"

"Sabina and Machele used color to label the classes and we used words," says Mario.

"Sabina and Machele's middle section has a lot of different arrays. It looks like a puzzle. Adoni, Adam, and Mario's middle section have arrays that are all the same length. They're all full rows of 10 seats long," Packmo notices.

Ms. Buchholz facilitates the discussion.

> Dillon noticed that everyone used partial products as a strategy for working with large numbers. But that does not mean everyone used the distributive property. Adoni, Adam, and Mario used the distributive property—their arrays are all the same length of 10 seats. You can see this in their equations to show that the sum of the small arrays is equal to the large array. This 10 remains the same. It is being distributed to each of the arrays. Sabina and Machele used partial products but not the distributive property. There is not a common factor in all of the partial products of their equation. With a partner, look at each group's work and find evidence of whether

Teaching Takeaway

We should use students' questions and responses to make adjustments to instruction.

they used partial products and then decide if they used the distributive property.

The students talk and analyze. Ms. Buchholz records the big idea that using partial products does not necessarily mean using the distributive property and listens to their conversations.

> Let's try to answer our essential questions for the day: How can we use what we know about multiplying and dividing small numbers to multiply and divide large numbers? How can we apply the properties (associative, commutative, and distributive) to problems with large numbers? Look across the shared work and our annotations. How would you answer these questions? Talk with a partner.

After 2 minutes, Ms. Buchholz has students share as she records their ideas. Decomposing into partial products is the strategy that comes up the most, so Ms. Buchholz stars this each time a student mentions it in his or her response.

Ms. Buchholz moves the class into the final phase of Bansho—the practice problems.

> EFFECT SIZE
> FOR DELIBERATE
> PRACTICE = 0.79

> We're going to practice applying these big ideas and strategies to problems without a context. I'm going to show you three equations. Two demonstrate applying the distributive property accurately to find the sum of partial products. One does not. You and a partner should talk and decide which two are true and which is a lie.

Two Truths and a Lie is a favorite game of the class and an engaging way to deliberately practice a transferred skill in a short period of time. Ms. Buchholz displays three equations and immediately her students are constructing arguments about why each is true or false. Some students pull out whiteboards or calculators. After 3 minutes, the class shares which equation is the lie and explains why.

To initiate closure and to make sure each student is accountable for the learning intentions, Ms. Buchholz transitions to the final reflection.

> EFFECT SIZE FOR
> EVALUATION AND
> REFLECTION = 0.75

Teaching Takeaway

Monitoring their own progress, seeking feedback, and being aware of their current level of understanding are characteristics of an assessment-capable visible learner.

EFFECT SIZE FOR ASSESSMENT-CAPABLE VISIBLE LEARNERS = 1.33

Video 4
Using Self-Reflection to Make Learning Visible

https://resources.corwin.com/ vlmathematics-3-5

For each of the four success criteria, you need one sticky note—red, yellow, or green depending on where you are with demonstrating that criteria. Put your initials on the sticky note and record your evidence to support the color you chose. Place your sticky note on your work in your math binder or on the board from today. If you have work from a previous day, you can place a sticky note on that as well."

The students study their work from today and write on colored sticky notes. This is students' first step toward creating their math portfolios. They will use their portfolios to reflect on their long-term learning across the unit and to set goals for the next unit. Ms. Buchholz is excited for her students to see their growth and to identify next steps. In this way, her students become their own teachers, continually transferring and applying concepts and skills to the ever-expanding number system. Figure 2.1 shows how Ms. Buchholz made her planning visible so that she could then provide an engaging and rigorous learning experience for her learners.

Ms. Buchholz's Teaching for Clarity PLANNING GUIDE

ESTABLISHING PURPOSE

1 What are the key content standards I will focus on in this lesson?

Indiana Academic Standards

3.C.5. Multiply and divide within 100 using strategies, such as the relationship between multiplication and division (e.g., knowing that 8 × 5 = 40, one knows 40 ÷ 5 = 8), or properties of operations.

3.AT.3. Solve two-step real-world problems using the four operations of addition, subtraction, multiplication, and division (e.g., by using drawings and equations with a symbol for the unknown number to represent the problem).

Standards for Mathematical Practice:

- Construct viable arguments and critique the reasoning of others.

- Look for and make use of structure.

2 What are the learning intentions (the goal and *why* of learning stated in student-friendly language) I will focus on in this lesson?

- Content: I am learning how strategies for multiplying and dividing small numbers can be transferred and revised to multiply and divide large numbers.

- Language: I am learning how mathematical properties can be used to describe and defend multiplication and division strategies.

- Social: I am learning to appreciate the contributions of each learner and the connections among others' reasoning and my own.

3 When will I introduce and reinforce the learning intention(s) so that students understand it, see the relevance, connect it to previous learning, and can clearly communicate it themselves?

- Align with transfer essential questions
- Introduce learning intentions and success criteria after activating prior knowledge about partial products
- Complete an observation checklist with success criteria
- Use colored sticky notes to document individual evidence

SUCCESS CRITERIA

4 What evidence shows that students have mastered the learning intention(s)? What criteria will I use?

I can statements:

- I can visually represent partial products in an array.
- I can decompose and compose products in multidigit multiplication using partial products.
- I can apply the distributive, commutative, and associative properties to multiplication of large numbers.
- I can transfer and revise efficient strategies for multiplication and division of large numbers.

5 How will I check students' understanding (assess learning) during instruction and make accommodations?

Formative Assessment Strategies:

- Observation/conference checklist with a list of anticipated strategies, success criteria, and planned questions
- Student work
- Sticky note evidence in math binders

Differentiation Strategies:

- Differentiate the content and product by readiness: open question
- Differentiate the process by situational interest: choose to work alone, with a partner, or in a small group. Choose materials.

INSTRUCTION

6 What activities and tasks will move students forward in their learning?

- Partial products number talk
- Auditorium task
- Two Truths and a Lie practice problems
- Sticky note evidence in math binders

7 What resources (materials and sentence frames) are needed?

Partial products number talk

Auditorium task

Two Truths and a Lie equations

Anchor charts of multiplication and division strategies

Graph paper

Open number lines and whiteboard markers

Colored pencils

Calculators

Colored tiles

Array cards

Student-made multiples number charts

Mathematical toolboxes

Colored sticky notes

Math binders

8

How will I organize and facilitate the learning? What questions will I ask? How will I initiate closure?

Instructional Strategies:

- Think-pair-share
- Number talk
- Bansho

Scaffolding Questions:

- How could you represent the size of each class using arrays and equations?
- How does rotating this array affect your solution?
- Why are these small arrays equal to this one large array? How could you represent this equality?
- How are you using partial products?

Extending Questions:

- How are you using other multiplication strategies from our anchor chart (making friendly numbers, skip counting, repeated addition)?
- Where do you see the associative property? Commutative property? Distributive property?
- How are you thinking about both multiplication and division in this problem?

Connecting Questions:

- What do you notice is the same across all the strategies and representations?
- How is _____ different from _____?
- How can we use what we know about multiplying and dividing small numbers to multiply and divide large numbers?

- How can we apply the properties (associative, commutative, and distributive) to problems with large numbers?

Self-Reflection and Self-Evaluation for Closure:

- Sticky note evidence in math binder

 This lesson plan is available for download at **resources.corwin.com/vlmathematics-3-5**.

Figure 2.1   Ms. Buchholz's Application Lesson on the Relationship Between Multiplication and Division

Ms. Mills and Equivalent Fractions and Decimals

For Ms. Mills, the beauty of mathematics lies in both its creativity and its omnipresence in life. She wants her fourth grade students to see the elegance of mathematics. Application tasks provide an opportunity for students to experience the value of their creative problem solving and the usefulness of their quantitative reasoning. Throughout each unit, Ms. Mills has students who explore the transfer level of learning by questioning and examining how to apply concepts and strategies to new contexts. For today, she designed a Color Run racecourse task to move all students forward to apply their fractional concepts and thinking skills in the new context of decimals.

> EFFECT SIZE
> FOR RELATING
> CREATIVITY TO
> ACHIEVEMENT
> = 0.40

> EFFECT SIZE
> FOR TRANSFER
> STRATEGIES = 0.86

COLOR RUN TASK

You will design and create a map of our school's 5-kilometer color run. Each of your team members holds two requirements to contribute to your map:

- Team Member 1: A color powder station is located every $\frac{1}{10}$ kilometer. A water station is located every $\frac{1}{2}$ kilometer.
- Team Member 2: Kilometer signs (in decimals) are located every $\frac{1}{15}$ kilometer. A bathroom is available every $\frac{3}{4}$ kilometer.

(Continued)

Teaching Takeaway

Providing learners with an authentic context promotes the application of concepts and thinking.

(Continued)

- Team Member 3: A water station is located every $\frac{2}{3}$ kilometer. You and your fellow cartographers may choose another area runners may need or want every $\frac{4}{5}$ kilometer.

Be sure to create a scale that shows the number of centimeters equivalent to what part of a kilometer on your map. Also create a key to show the symbol for each station. Your final product should also include a chart to show the number of each station you will need for the full color run race. This race is a loop. Be sure to mark the start and finish line.

online resources — This task is available for download at **resources.corwin.com/vlmathematics-3-5**.

EFFECT SIZE FOR PERCEIVED TASK VALUE = **0.46**

EFFECT SIZE FOR PRIOR ACHIEVEMENT = **0.55**

Most recently, her students expanded their conceptual understanding and procedural knowledge of addition and subtraction to fractions with unlike denominators. Her students fluently move between fraction equivalents and use their fraction number sense to estimate computations and then check solutions' reasonableness. In an earlier place value unit, Ms. Mills and her students extended the base-ten number system by attaching decimal notation to the language and representations of tenths, hundredths, and thousandths.

Today, Ms. Mills is transitioning students into the next unit on fraction and decimal equivalents as well as decimal computation within the context of the metric system. The big ideas of the new unit are as follows: *Fractions and decimals are equivalent (equal) if they are the same size or the same point on a number line; a fraction with denominators 10, 100, or 1,000 can be represented by an equivalent decimal. Equivalent measures can be expressed within and across measurement systems; the metric system relies on place value.* Ms. Mills wants her students to bring their deep conceptual understanding and procedural knowledge of fractions and transfer it to the realm of decimals to deepen their surface learning of decimals. This application will solidify her students' fraction number sense while further developing their proportional reasoning and appreciation of the elegance of mathematics.

EFFECT SIZE FOR CLEAR GOAL INTENTIONS = **0.48**

EFFECT SIZE FOR STRATEGY TO INTEGRATE WITH PRIOR KNOWLEDGE = **0.93**

What Ms. Mills Wants Her Students to Learn

There are many possible sequences of topics, and Ms. Mills relies on her division pacing guide to provide the scope and sequence of mathematical units in fourth grade. From this big picture view, Ms. Mills finds the meaningful connections and the ways each unit is a building block toward the next. Then she selects and creates application tasks with a low mathematical floor and a high mathematical ceiling (Boaler, 2016) to make this connection visible to students. In this way, her students engage in transfer-level learning with fraction equivalents and surface-level learning with decimal measurements as she connects fractions with decimals.

In today's lesson, Ms. Mills's students will begin to examine the following.

> ### VIRGINIA MATHEMATICS STANDARDS OF LEARNING
>
> 4.3. The student will (a) read, write, represent, and identify decimals expressed through thousandths; (c) compare and order decimals; and (d) given a model, write the decimal and fraction equivalents.
>
> **Ms. Mills is helping her learners develop the following Mathematical Process Standards:**
>
> - Mathematical connections
> - Mathematical problem solving
> - Mathematical representations

Learning Intentions and Success Criteria

Ms. Mills creates learning intentions based on the content and process standards. She also uses the learning intentions to communicate their

Teaching Takeaway

The application of concepts and thinking requires learners to detect similarities and differences between concepts, thinking, and situations.

Teaching Takeaway

Low-floor/high-ceiling tasks ensure all learners have access to the application of concepts and thinking.

goal of applying fraction concepts and thinking skills to decimals. She knows being transparent and explicit will allow her students to make meaningful connections between what they already know and what they are learning. Ms. Mills creates three types of learning intentions to highlight the mathematical content, the mathematical language, and the social interactions at the core of students' learning:

Content Learning Intention: I am learning to understand the relationship between equivalent fractions and decimals.

Language Learning Intention: I am learning to understand the language of equivalent fractions and decimals on a length/linear model

Social Learning Intention: I am learning to understand how to persevere and attend to precision by relying on each other's mathematical reasoning and questioning.

EFFECT SIZE FOR COGNITIVE TASK ANALYSIS = 1.29

Ms. Mills creates success criteria because she wants her students to be able to describe themselves in relation to the learning intentions. The success criteria are precise, while allowing for students' entry at the surface, deep, and transfer levels of learning:

- ☐ I can identify equivalent fractions and decimals and count by fractions and decimals.
- ☐ I can precisely measure and label fraction and decimal distances on a number line.
- ☐ I can name benchmark fractions and their equivalent decimal benchmarks.
- ☐ I can use fraction, decimal, and place value language to describe the equivalent values.

EFFECT SIZE FOR TEACHER CLARITY = 0.75

Her students will engage in making sense of the learning intentions and success criteria before they begin the Color Run task. As a class, they will use the success criteria to evaluate a worked example of a similar task. In this way, Ms. Mills ensures her students have a vision for processes and products that demonstrate mastery of the learning intentions.

Activating Prior Knowledge

Ms. Mills's students slowly put their thumbs up, hold them at chest level while sitting on the carpet in the middle of her room, and look at a now-dark screen. The thumbs up lets Ms. Mills know her learners are ready. They are engaged in a math routine called quick images. Quick images are flashed images of quantities that are organized in a way that makes them easy to subitize or quickly recognize without counting (Clements & Sarama, 2007). Ms. Mills says, "Remember, the 10 × 10 grid is one whole. I will flash the image again so you can revise or confirm your thinking in three, two, one." The screen fills with an image of a 10 × 10 grid where five out of ten columns are shaded.

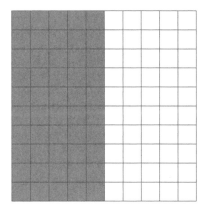

After 3 seconds, the image disappears again. More thumbs are up. The teacher says, "Turn and tell a partner how much of the grid was shaded and how you know."

After 1 minute of talk, Ms. Mills asks for someone to share how they saw the quick image or how their partner saw it. "I saw half of the 10 × 10 grid shaded. It's one-half or five-tenths," says Nivek.

"How could I write that?" Ms. Mills asks.

Nivek responds, "You could write either fraction: $\frac{1}{2}$ or $\frac{5}{10}$. They're equivalent."

Ms. Mills writes the two fractions and asks, "Now, what if I wanted to write it as a decimal?"

"You could write five-tenths as 0.5. That's also five-tenths," Keshira adds.

Teaching Takeaway

Questioning is an effective way to make student thinking visible.

Video 5
Teaching Reflection Skills
Starts With Clear
Learning Intentions and
Success Criteria

*https://resources.corwin.com/
vlmathematics-3-5*

Ms. Mills records $\frac{1}{2} = \frac{5}{10} = 0.5$ and then continues with the next two quick images: a 10×10 grid showing $\frac{1}{4}$ or 0.25 and one showing $\frac{1}{10}$ or 0.1.

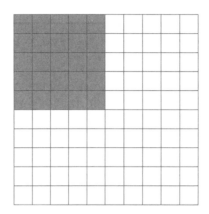
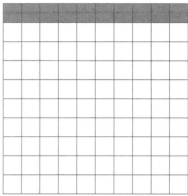

Similarly, she flashes each image for 3 seconds, gives about 20 seconds of think time, and then reflashes the image so students can confirm or revise their thinking. Students think-pair-share how much of the grid was shaded and how they knew. Ms. Mills records equivalencies, emphasizes equivalent decimals, and then adds new information to the class-created anchor charts. Through this math routine, the students' prior knowledge related to decimal and fraction language, notation, and visual models is activated. The students will transfer this procedural knowledge to the context of equivalent fractions and decimals and to linear models today. Now, they are ready to examine today's application task.

EFFECT SIZE FOR
IMAGERY = 0.45

Ms. Mills spends time engaging in class discussion to be sure the task is understood and the expectations are clear (Van de Walle, Karp, & Bay-Williams, 2018).

> As you know, our color run race is coming up and the race planners are making decisions about the racecourse. They have decided the race will be a loop. They need several stations along the route: the color powder, water, bathrooms, and distance markers. Today, you're going to work in teams of three to create a suggested racecourse. Each of your team

members will bring certain requirements that you have to implement.

Our learning intentions are focused on taking what you already know about fractions, fraction models, fraction language, and fraction equivalencies and applying that knowledge to decimals. Let's read the learning intentions.

Ms. Mills asks students to think-pair-share what they notice and what they wonder about the learning intentions.

When they share with the whole group, William wonders what a length or linear model is. Lucas points to several models on anchor charts and explains, "It could be a ruler, number line, Cuisenaire rods, or fraction bars. Really anything that shows length or measures length."

William makes the connection: "Oh, like a racecourse."

Next, Ms. Mills shares the success criteria. "To get ready for our work today, we're going to look at last year's racecourse and evaluate it using today's success criteria. Here is last year's course. Take a moment to study it and think about what you notice and what you wonder." Ms. Mills waits 2 minutes. The wait time conveys that everyone should be noticing and wondering thoughtfully, not just superficially. She continues, "Now turn and talk with a partner about what you notice and what you wonder." Students turn and share animatedly with a partner. A visual image (Figure 2.2 on the next page) paired with the open questions of noticings and wonderings allows for every student to access the task, whether they are at the surface, deep, or transfer levels of learning. After 1 minute, each pair shares one noticing or wondering with the whole class.

Ms. Mills continues, saying "Now, let's look at each success criteria and evaluate this course. The first criterion states, *I can identify equivalent fractions and decimals and count by fractions and decimals.* Where do you see evidence of this?"

Ava begins, "I see evidence of counting by halves: $\frac{1}{2}$, $\frac{2}{2}$, $\frac{3}{2}$, $\frac{4}{2}$. Those are the distance markers for the miles."

Video 6
Consolidating Prior Learning Before Starting an Application Task

https://resources.corwin.com/vlmathematics-3-5

Teaching Takeaway

Wait time increases student engagement.

EFFECT SIZE FOR CLASSROOM DISCUSSION = **0.82**

WORKED EXAMPLE OF A 2-MILE RACECOURSE

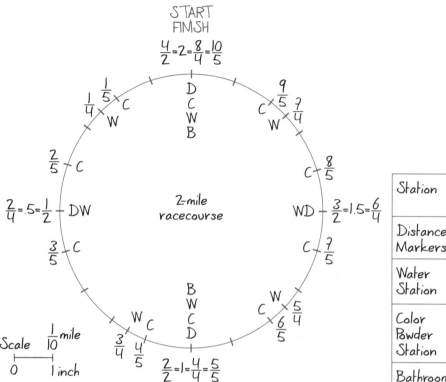

Figure 2.2

"They have equivalent fractions marked. At two-halves, they have 1 mile. At four-halves, they have 2 miles," Benjamin continues.

Charles adds, "I see five-tenths written as a decimal. It's equivalent to $\frac{1}{2}$, just like in the quick image. And then three-halves is also one and five-tenths. So that shows they identified equivalent fractions *and* decimals."

"I want to go back to what Benjamin said. I think that what he said combined with Charles' noticing shows they can count by decimals. It starts five-tenths, then one, then one and five-tenths, and then two. That's counting by five-tenths," Leah points out.

Kamora says, "I agree with Leah. They're counting by halves. One-half in decimals is five-tenths."

Jayden interjects, "I disagree with Leah and Kamora. I think to show they can count by decimals, they would have to count by a smaller number. Like if they counted the distance markers by decimals instead of just by fractions." He points to the water stations at every quarter mile. "This would be . . . point two five?" Jayden looks at the anchor chart where they added $\frac{1}{4} = 0.25$ during the quick images routine.

"How do you say that decimal mathematically?" Ms. Mills poses to the class.

"Twenty-five hundredths," Calvin replies.

Jayden continues his explanation.

> So this would be 25 hundredths and then this would be five-tenths and then . . . well . . . I'm not sure. But I know you could say another decimal instead of three-fourths. It would go here because it would be equivalent to three-fourths. That would be evidence that you can count by decimals.

Jayden points to the water station on the course race that shows three-fourths.

Ms. Mills decides to highlight this point.

> Jayden is saying we need more evidence to be sure this mathematician can count by decimals. He is arguing that counting by a smaller decimal will show more decimal knowledge. That's something to consider as you create your racecourse. How much work with decimals do you need to show in order to demonstrate mastery of the success criteria?

Teaching Takeaway

Classroom dialogue, either through questioning or discussion strategies, provides feedback to the teacher about student thinking and learning.

Teaching Takeaway

Using student ideas to determine the next steps in the learning process is essential in the Visible Learning classroom.

Should you use every opportunity to show what you know about decimals or should you only show what is comfortable and easy for you?

Saw offers the following explanation:

> Well if it's comfortable and easy, then we already know it. But I don't think we already know everything about equivalent fractions and decimals or we wouldn't be working on it today.

> So we should work on showing decimals that are hard for us. Like when Jayden was trying to figure out the decimal for three-fourths. We should work on that because three-fourths is a benchmark fraction and the third success criterion says, *I can name benchmark fractions and their equivalent decimal benchmarks.*

With this comment, Ms. Mills begins to transition the class to discussing the third success criterion. She reminds students of the second success criterion (*I can precisely measure and label fraction and decimal distances on a number line*) and then asks, "What if one of the requirements is to include a safety patrol at every one-tenth of a mile? How could I add this to the racecourse?" Several students take turns analyzing the steps of the task and modeling how to use a ruler to create one-tenth intervals based on the scale. Then they add labels to the racecourse and key. Eventually, the class has unpacked each success criterion by evaluating the worked example.

Before she sends students to start their group work, Ms. Mills points out a difference between the worked example and today's task: "The race planners are making one big change: Last year, the color run was a 2-mile race; this year, it will be a 5-kilometer race." She then reminds her students of today's task and the roles of the three team members.

> This task will take us a few days. At the end of our work time today, we will practice evaluating one team's course in order to give them feedback for revisions and additions. Tomorrow, we will continue working on this task. Remember the resources

Teaching Takeaway

Worked examples allow learners to see what mastery looks like for the given success criteria. This, in turn, supports learners as they engage in strategy monitoring, evaluation, and reflection.

you have available to help you with problem solving and modeling.

Ms. Mills points to the manipulatives, graph paper, calculators, colored pencils, and anchor charts and continues.

Also remember that this task will take perseverance and attention to precision. Rely on your team members' mathematical reasoning and questioning, and check in with the success criteria and last year's course for support. I will also be meeting with teams as you work in order to hear about your process and thinking.

Scaffolding, Extending, and Assessing Student Thinking

Ms. Mills has purposefully created the teams using a heterogeneous grouping strategy called alternate group ranking (Frey et al., 2018). To create teams of three students, Ms. Mills put a recent self-evaluation of unit learning intentions in descending order from highest score to lowest and then sorted them, one at a time, into seven stacks. This resulted in the groups shown in Figure 2.3.

ALTERNATE GROUP RANKING

Team 1	Team 2	Team 3	Team 4	Team 5	Team 6	Team 7
Student 1	Student 2	Student 3	Student 4	Student 5	Student 6	Student 7
Student 8	Student 9	Student 10	Student 11	Student 12	Student 13	Student 14
Student 15	Student 16	Student 17	Student 18	Student 19	Student 20	Student 21

Figure 2.3

Ms. Mills relies on these groups as a starting point; they reflect heterogenous *current* abilities. Then she makes adjustments based on the language needs of her students, specific pairings she knows to be productive, and social concerns. Her goal is to create teams where no two students are "too far apart" to communicate productively and where they can focus on the social learning intention of persevering and attending to precision by relying on each other's mathematical reasoning and questioning. Her creation of the small group task also reflects this learning intention because each team member holds two unique requirements, which necessitates that team members communicate and collaborate in order to include each requirement.

Ms. Mills passes out the main task and requirements to team members. Students check the team rosters on the board, find each other, and settle into work spaces. Ms. Mills collects her clipboard with her conference chart for today's task. Her conference chart is a checklist of the four success criteria with space for notes (Figure 2.4). Students on the same team are clustered together but each student has his or her own row. This chart will allow Ms. Mills to check for evidence of individual student learning and to note trends within and across teams.

Ms. Mills also uses her conference chart as a resource as she confers with students. She includes a list of materials to scaffold students who are struggling and her list of planned questions to extend and assess student thinking. Before teaching the lesson, Ms. Mills anticipates student strategies. Anticipating enables Ms. Mills to plan her conference chart to monitoring student learning and to identify scaffolding strategies, materials, and questions that will support students without removing them from productive struggle (Smith & Stein, 2011). Ms. Mills knows that effective, differentiated instruction requires intentional planning. The time spent planning in advance of the lesson maximizes her instructional time and students' growth toward the learning intentions.

The teams work in spaces spread around the classroom. Students have a stack of three cups near them to nonverbally signal to Ms. Mills how they are doing: A red cup on top means they are stuck and cannot move forward until they confer with Ms. Mills. A yellow cup on top

EFFECT SIZE FOR
ABILITY GROUPING
= 0.12

EFFECT SIZE
FOR PROVIDING
FORMATIVE
EVALUATION = 0.48

Teaching Takeaway

As teachers, we must ensure that we are collecting evidence of learning for individual students so that we can make adjustments to our teaching.

EFFECT SIZE FOR
RECORD KEEPING
= 0.52

OBSERVATION/CONFERENCE CHART

Date _____

Observation/Conference Checklist

Name	SC 1: I can identify equivalent fractions and decimals and count by fractions and decimals.	SC 2: I can precisely measure and label fraction and decimal distances on a number line.	SC 3: I can name benchmark fractions and their equivalent decimal benchmarks.	SC 4: I can use fraction, decimal, and place value language to describe the equivalent values.	Notes

Overall Patterns:

Questions:

- What is an equivalent fraction and/or decimal? How do you know?
- What would this fraction/decimal look like on a 10 × 10 grid?
- How many stations will there be per kilometer?
- How many meters is 0.15 km?
- What if the track was 10 km long? What if the track was out-and-back? How would the station locations change?

Materials:

- 10 × 10 grid paper
- Base-ten blocks
- Cuisenaire rods
- Fraction bars
- Whiteboard number lines
- Calculators

Figure 2.4

 This template is available for download at **resources.corwin.com/vlmathematics-3-5**.

Teaching Takeaway

Teaching mathematics in the Visible Learning classroom involves getting feedback from our learners about where to go next.

Teaching Takeaway

Closure is an important aspect of a learning experience. Closure helps to consolidate student learning.

means they have a significant question or conflict but they are able to work on something else productively until Ms. Mills can confer. A green cup on top means they are working productively without immediate need for a conference. Ms. Mills scans the room as she finishes conferring with a team to note who needs to meet with her and at what level of priority.

During the 40 minutes of work time, Ms. Mills is able to confer with four of seven teams. She identifies one team to share their in-progress racecourse design during the whole-group share today. She also notes an overall pattern in students' confusion: *What do you do when the intervals don't make 5 kilometers exactly, like the water stations?* This will guide her mini-lesson focus for tomorrow. Ms. Mills also noted the three teams she needs to confer with tomorrow.

Teaching for Clarity at the Close

When Ms. Mills sounds her chime, students clean up. Ms. Mills highly values this sharing time and works to include it each day. This is when her class comes together as a learning community of mathematicians to share and celebrate their successes, mistakes, and questions. She strategically chooses when to focus on the content, craft, process, or progress in students' sharing (Thunder & Demchak, 2012). Based on her anticipation of strategies and mistakes, Ms. Mills makes plans to monitor, select, and sequence specific content, craft, process, or progress aligned with the day's learning intentions (Smith & Stein, 2011). She also plans questions to help students make meaningful connections among what is shared.

Similar to the mini-lesson, one team shares their racecourse draft and the class discusses each success criterion, looking for "Gems and Opportunities." As a community, they look for evidence of mastery to celebrate (gems) and offer suggestions for revisions and additions (opportunities). Ms. Mills records each on sticky notes, modeling how students will provide feedback via gems and opportunities independently during a future lesson. This process not only provides the sharing team with timely, specific feedback aligned to the success criteria, but it also gives all teams an opportunity to reflect on their work through comparison and analysis.

After the whole-group discussion, Ms. Mills gives each team a two-column chart with the following headings: Gems (Great! Keep doing this!) and Opportunities (Make a change. Be sure to add.).

> From evaluating Charles, Zahara, and Keshira's racecourse draft, you should be thinking about your own draft and what you did well and what you need to revise. Take 3 minutes to discuss this with your team and make notes of gems and opportunities for your own racecourse draft based on the success criteria. You will use these notes as a plan for your work tomorrow.

Each team talks, looks at their draft, refers to the success criteria, and records their gems and opportunities.

The teacher says, "Now, flip over your Gems and Opportunities chart. There are three questions for you to discuss with your team. As you discuss, take turns recording your responses. Be specific." The three questions are as follows:

1. Compare your scale to the sharing team's scale. What unit represents what fraction of a kilometer? How did this help you to be precise in your measurements?

2. Make a list of benchmark fractions that you labeled and their equivalent decimal benchmarks. How could you prove these are equivalent?

3. Think about the fraction, decimal, and place value language you used while working. How is the place value language connected to the decimal and fraction language? Why?

As teams discuss, Ms. Mills joins each group and makes additional notes on her conference chart. She has intentionally sequenced tasks that develop conceptual understanding before procedural knowledge and fluency. She has engaged her students in regular, spaced practice of counting and subitizing rational numbers (Thunder & Demchak, 2016). Now, her students are successfully applying these concepts and skills to the new context of decimals. She will collect their work and use it to plan her mini-lesson, conference questions, and sharing tomorrow. This student work will also guide her planning and differentiation of other tasks within the decimal unit as she analyzes who is at the surface, deep, and transfer levels of learning related to decimals.

Ms. Mills notices some students flipping back to the gems and opportunities side of the paper to make additional notes. This gives Ms. Mills important feedback about her questions; these self-monitoring and analysis questions are successfully making students' learning visible to herself and to the students themselves. Figure 2.5 shows how Ms. Mills made her planning visible so that she could then provide an engaging and rigorous learning experience for her learners.

Ms. Mills's Teaching for Clarity PLANNING GUIDE

ESTABLISHING PURPOSE

1 What are the key content standards I will focus on in this lesson?

Virginia Mathematics Standards of Learning

4.3. The student will (a) read, write, represent, and identify decimals expressed through thousandths; (c) compare and order decimals; and (d) given a model, write the decimal and fraction equivalents.

Mathematical Process Standards:

- Mathematical connections
- Mathematical problem solving
- Mathematical representations

2 What are the learning intentions (the goal and *why* of learning stated in student-friendly language) I will focus on in this lesson?

- Content: I am learning to understand the relationship between equivalent fractions and decimals.
- Language: I am learning to understand the language of equivalent fractions and decimals on a length/linear model.
- Social: I am learning to understand how to persevere and attend to precision by relying on each other's mathematical reasoning and questioning.

3 When will I introduce and reinforce the learning intention(s) so that students understand it, see the relevance, connect it to previous learning, and can clearly communicate it themselves?

- Post learning intentions
- Notice and wonder about learning intentions
- Make connections during worked example evaluation, conferences, and sharing
- Gems and Opportunities team reflection

SUCCESS CRITERIA

4 What evidence shows that students have mastered the learning intention(s)?
What criteria will I use?

I can statements:

- I can identify equivalent fractions and decimals and count by fractions and decimals.
- I can precisely measure and label fraction and decimal distances on a number line.
- I can name benchmark fractions and their equivalent decimal benchmarks.
- I can use fraction, decimal, and place value language to describe the equivalent values.

5 How will I check students' understanding (assess learning) during instruction and make accommodations?

Formative Assessment Strategies:

- Success criteria conference checklist
- Racecourse draft
- Gems and Opportunities team reflection

Differentiation Strategies:

- Differentiate the content and product by interest: Create a scale. Create a sixth station.
- Differentiate the process by readiness: Create small groups using alternate group ranking.

INSTRUCTION

6 What activities and tasks will move students forward in their learning?

- Quick images
- Worked example evaluation

- Modeling
- Color Run small group task
- Gems and Opportunities draft sharing
- Gems and Opportunities team reflection

7 What resources (materials and sentence frames) are needed?

Quick images

Worked example (last year's Color Run course)

Anchor charts of equivalencies

Language frames

Cuisenaire rods

Fraction bars

Base-ten blocks

10×10 grids

Graph paper

Open number lines and whiteboard markers

Colored pencils

Calculators

8 How will I organize and facilitate the learning? What questions will I ask? How will I initiate closure?

Instructional Strategies:

- Worked example
- Self-evaluation
- Anticipate, monitor, select sequence, and connect students' strategies
- Turn and talk

Scaffolding Questions:

- What is an equivalent fraction and/or decimal? How do you know?
- What would this fraction/decimal look like on a 10 × 10 grid?

Extending Questions:

- How many stations will there be per kilometer?
- How many meters is 0.15 km?
- What if the track was 10 km long?
- What if the track was out-and-back? How would the station locations change?

Connecting Questions:

- Compare your scale to the sharing team's scale. What unit represents what fraction of a kilometer? How did this help you to be precise in your measurements?
- Make a list of benchmark fractions that you labeled and their equivalent decimal benchmarks. How could you prove these are equivalent?
- Think about the fraction, decimal, and place value language you used while working. How is the place value language connected to the decimal and fraction language? Why?

Self-Reflection and Self-Evaluation for Closure:

- Gems and Opportunities team reflection

 This lesson plan is available for download at **resources.corwin.com/vlmathematics-3-5.**

Figure 2.5   Ms. Mills's Application Lesson on Equivalent Fractions and Decimals

Ms. Campbell and the Packing Problem

Ms. Campbell is pleased by the progress her fifth grade students have made in learning about volume. They have seen connections between volume and area, and they have shown good algebraic thinking in their work with the formula. It is time to put this learning into practice for the real world.

Ms. Campbell wants students to use their knowledge of volume to solve a problem about a topic that is relevant, engaging, and important to them. She created this task because she knows her students enjoy riding scooters at recess. She also knows that their parents insist that they use helmets whenever they ride their scooters. Thus, she thinks scooters and helmets would be the perfect topic for them to extend their thinking about volume.

In addition, Ms. Campell also knows that scooters and helmets are most always shipped in different size boxes from the manufacturers. Finally, she realizes that working with different size boxes will challenge her students to work beyond their knowledge of the volume formula to reason about volume when combining objects of different shapes. Now, the elements to pack are the boxes containing scooters and helmets, and the shapes are not the unit cubes students have grown accustomed to working with over the past several days.

Ms. Campell begins today's lesson as follows:

> Class, we are going to solve a packing problem today for the end of our unit on volume. The school is going to purchase some scooters with helmets. Each scooter and each helmet comes inside its own box, and then the order will be shipped in larger shipping boxes. Our job is to figure out how many boxes it will take to ship them here, so we can be sure to have enough space in the storage room.

PACKING CHALLENGE

Your school has ordered scooters and helmets for use during recess. Each scooter ships folded in a box that measures 1' × 1' × 3' and weighs 5 pounds. Each helmet ships in a box measuring 0.5' × 1' × 1' and weighs 3 pounds.

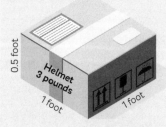

The order includes 14 scooters and 14 helmets. The scooters and helmets will be bundled together into larger shipping boxes. Those boxes are 2' × 2' × 3'. How many shipping boxes will it take to ship the school's order? What is the best way to pack the order? Explain your reasoning using pictures, equations, and words.

Sources: FamVeld/iStock.com (child on scooter) and aurielaki/iStock.com (boxes)

What Ms. Campbell Wants Her Students to Learn

Ms. Campbell wants this lesson to help her students put their knowledge of volume into action. She has chosen to focus this assignment on the standard that emphasizes that volume is additive. Students must figure out how the volume of each scooter or helmet box adds together and fits into the shipping boxes.

EFFECT SIZE FOR STRATEGY TO INTEGRATE WITH PRIOR KNOWLEDGE = 0.93

MATHEMATICS CONTENT AND PRACTICE STANDARDS

5.MD.C. Understand concepts of volume and relate volume to multiplication and to addition.

5. Relate volume to the operations of multiplication and addition and solve real-world and mathematical problems involving volume.

c. Recognize volume as additive. Find volumes of solid figures composed of two non-overlapping right rectangular prisms by adding the volumes of the non-overlapping parts, applying this technique to solve real-world problems.

Ms. Campbell is helping her learners develop the following Standards for Mathematical Practice:

- Model with mathematics.
- Attend to precision.

Students will model with mathematics when they use mathematics to quantify, describe, and solve the packing problem. They will use math as a tool to help them solve a problem in their world. They will attend to precision as they give a clear description of their solution and their rationale for packing the boxes as they have.

Learning Intentions and Success Criteria

Ms. Campbell realizes that her students must have a clear picture of what is expected of them in each lesson. She uses learning intentions

EFFECT SIZE FOR
TEACHER CLARITY
= 0.75

and success criteria to show her students what is important in today's lesson and what success with the learning experience will look like.

> *Content Learning Intention*: I am learning how to solve a problem about packing based on what I know about volume.
>
> *Language Learning Intention*: I am learning to describe the solution precisely using the language of volume and the language of the problem situation.
>
> *Social Learning Intention*: I am learning how to share my thinking with my partner and listen to the ideas my partner has about the problem.

EFFECT SIZE FOR
COGNITIVE TASK
ANALYSIS = 1.29

Ms. Campbell creates success criteria so that her students can monitor their own progress in relation to the learning intentions. The success criteria give students precise skills or tasks they should be able to complete if they have mastered the learning intentions.

> ☐ I can tell my partner my ideas about how to solve the problem.
> ☐ I can find the volume of each item to be packed and use that to solve the problem.
> ☐ I can add volumes to find the total volume of a group of shapes.
> ☐ I can explain why my solution to the problem is a good one.

Activating Prior Knowledge

Ms. Campbell is excited to see how her students respond to this challenge. It is early in the year and the students have not yet had much experience solving problems like this with her. She is confident that they understand the volume formula and will be able to find the volume of right rectangular prisms given the dimensions. This task asks her students to think creatively to figure out how to pack these less familiar shapes into larger packing boxes.

Because the problem is complex, Ms. Campbell begins the lesson by allowing time for students to discuss the problem, first with partners and then as a class. She distributes the problem to the class and asks them to talk with their shoulder partner about the first paragraph of the problem.

Ms. Campbell says, "As you read and discuss the information, see if the diagrams can help you make sense of the information in the problem. Think about what information you do understand and what questions you have, which may help with things you do not understand." As the class talks, she moves among the groups, listening and asking focusing questions. Each group talks clearly about two items—scooters and helmets—and understands that the boxes containing each are different sizes. The notation for volume seems to be the challenge. Ms. Campbell anticipated this and is ready with questions to connect the different representations.

"I'm not sure what the apostrophe means in math class." Alex says to Ms. Campbell when she approaches his table. "If the apostrophe were not there, what would you think about this expression?" Ms Campbell asks, pointing to $1' \times 1' \times 3'$ in the problem. "That's one times one times three," Alex replies. Ms. Campbell asks, "How can the diagram help you figure out the units for those values?" She walks away, thinking about how to better introduce students to the symbols used to represent the volume of a prism. Symbols and notation are important, but she is worried that the students are dedicating problem-solving time to the task of deciphering symbols.

At another table, Cecelia recognizes the pattern of $1' \times 1' \times 3'$ as length times width times height, one of the volume formulas the class has been studying. Cecelia asks, "I don't know why the box is lying down in the picture. If three is the height, should it be tall?" Ms. Campbell hears her partner respond, "Remember when we built the prisms and put them on different sides? I think that's happening here. It's on its side."

When the class returns to a whole-group discussion, Ms. Campbell asks them to talk about what they understand and what questions they have. Alex says, "We figured out that the apostrophe means foot in math. The expression has the apostrophe but the picture has the word foot."

The teacher responds, "Thank you, Alex. Who has another observation to share?"

"We think the numbers are in a different order than the picture," Cecelia begins. "It's like the box is lying on its side. But because we're multiplying, that doesn't matter."

EFFECT SIZE FOR COGNITIVE TASK ANALYSIS = **1.29**

EFFECT SIZE FOR QUESTIONING = **0.48**

EFFECT SIZE FOR HELP SEEKING = **0.72**

EFFECT SIZE FOR CLASSROOM DISCUSSION = **0.82**

Ms. Campbell wonders if everyone sees what Cecelia is seeing about the orientation of the box. She knows that for some students, it is difficult to rotate the box in their minds and identify all of the dimensions. In fact, some students may still not see that there are three dimensions. She makes a note to encourage students next to bring out the linking cubes or to make models of the boxes with nets so that they can explore the dimensions with their hands.

For now, she takes advantage of the opportunity to identify an arithmetic property in a real-life context. "Who can remind us what property of multiplication Cecelia is talking about?" Ms. Campbell asks.

EFFECT SIZE FOR
FEEDBACK = 0.70

As the discussion continues, the class decides to write a key on the board to help them with the problem.

$$1' \times 1' \times 1' = \text{one foot by one foot by one foot}$$
$$= \text{length times width times height}$$

Ms. Campbell asks the students to (re)read the second part of the problem individually. She knows some groups looked at it during the earlier conversation and says, "You will see this notation again so check that you understand it. This is also the time to think about what new information is shared and what the question is asking. I'll ask for a volunteer to state the question in their own words after you read."

EFFECT SIZE FOR
SUMMARIZATION
= 0.79

As individuals finish reading, shoulder partners begin talking quietly. Ms. Campbell listens to the conversations among students as others finish reading and have a moment to think. She hears students asking each other questions like these: "Will it all fit in one box?" "Do you have to pack one scooter and one helmet together?" "What happens if the box is not full?" "No, I don't think you have to leave the box laying down on the long side. You can stand it up, can't you?"

EFFECT SIZE FOR
SMALL GROUP
LEARNING = 0.47

EFFECT SIZE
FOR STRATEGY
MONITORING = 0.58

After asking and answering each other's questions about the problem, the students set to work in table groups. Ms. Campbell knows that her students will use a variety of tools, so she has rulers, unit cubes, and paper for sketching available for her students. As students begin to work, she sees some groups use rulers to measure and model the shape of the scooter box base. Other groups use linking cubes to build models of the scooter box. Ms. Campbell asks about their work as she approaches the table: "Tell me about the model you've made here."

The students respond, "We have these three blocks to be the scooter box. It's one by one by three just like the models we built last week. This [single] one is two helmet boxes because we can't cut the blocks in half."

"Tell me more about that," Ms. Campbell says. She is wondering if the students have noticed that the helmet box is only 0.5 feet high and not 1 foot high like the scooter box, because their model shows the heights as the same on that dimension.

EFFECT SIZE FOR
ELABORATION AND
ORGANIZATION
= 0.75

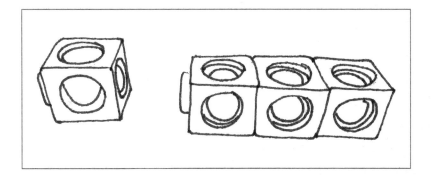

Judy explains her group's thinking about the model: "Each helmet box is one half of a foot high. If one side of the cube is 1 foot, then it takes two helmet boxes on top of each other to be 1 foot high like the cube. This is two boxes." Ms. Campbell sees the other students nod with pride at this clear explanation.

Ms. Campbell says, "I hear what you are saying, but I am having a hard time understanding something. How tall is the helmet box?"

The students point to the single cube.

"How tall is the scooter box?" the teacher asks.

The students point to a row of three green cubes. "Wait!" says Judy, "That can't be right! They're the same height."

Ms. Campbell responds by saying, "Can you draw a picture to show what you just said?" Because the students seem to have a new understanding of their concrete model, Ms. Campbell leaves the group with this question as she moves on to another group.

At another table, Ms. Campbell sees students discussing a sketch.

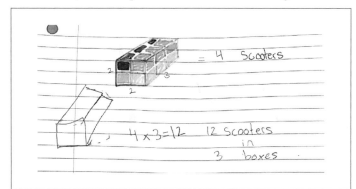

"Tell me about the colors in your diagram," Ms. Campbell says.

"Each color is one scooter. You can only see the front of this box [pointing to the bottom left square]. The big box holds four scooters," one student explains. "That means three big boxes hold 12 scooters," another student at the table continues. "We still have to figure out about the helmets and the other two scooters," says another student. Ms. Campbell moves on as the group goes back to discussing their work, but she makes a note to ask this group later to describe their strategy for packing the helmet boxes on her next trip back to their table. She notes that this is a good opening for discussing the pros and cons of packing the helmet boxes and scooter boxes separately.

Scaffolding, Extending, and Assessing Student Thinking

Before she taught the lesson, Ms. Campbell worked through the task herself. She wanted to anticipate the approaches her students could take and the problems they might have. As she anticipated, Ms. Campbell notices that the groups are packing all of the scooters and then all of the helmets. This means there is a last box with only two helmets and a lot of empty space! As teams finish recording their solutions, Ms. Campbell is ready with one more question to extend student thinking: "Is this the only way the boxes can be packed? Are all of these boxes the same weight?" The weight information had been ignored so far in the process as extra information. Ms. Campbell's question pushes the groups to think about weight as part of packing shipping boxes. William responds, "The box with four scooters weighs 20 pounds. Twelve helmets weighs

36 pounds, and then adding two scooters means this box is 46 pounds. That's heavy!" Ms. Campbell then asks, "What is that like for the delivery driver? Can you make the weight more even?"

This is what mathematical modeling is about. Students are using their math knowledge to describe and solve a problem in the real world by seeing the problem through a mathematical lens. As Ms. Campbell leaves the group to think about her question, she hears them talk about not putting so many helmets in one box. They are on their way to finding another (potentially better) solution to the problem.

> EFFECT SIZE FOR TRANSFER STRATEGIES = 0.86

Teaching for Clarity at the Close

Ms. Campbell says, "Tomorrow we will share our solutions to this problem, but before we stop, I'd like to ask each of you to write down what you know about the volume of the boxes and the arrangements you have so far." As students begin the task, she listens to their conversations. "The helmet box has a volume of 1!" "No, I think it is 0.5, because $1 \times 1 \times 0.5 = 0.5$. See? Look at what we wrote on the board!" "I agree. When we made the model with linking cubes we saw the same thing." "I wonder how much the shipping box holds?" Ms. Campbell is pleased to see that students are looking up at the board and using the key that they created and connecting it to the models, but there is still work to be done to make sense of a model of volume and the expression that is used to represent it. She addresses the students, saying "To end the lesson today, I want you to think about our success criteria." Ms. Campbell points to them on the board and asks the students to take out their math journals.

> Please rate your work on each criterion using our red/yellow/ green system and then write down one example of something good you did and a question you still have. After you and your partner both finish, share your examples with each other. Remember that you want to be sure you record questions for criteria you mark as yellow or red because we know you need to know more about those things.

> EFFECT SIZE FOR EVALUATION AND REFLECTION = 0.75

Ms. Campbell has found that her students are more thoughtful about their ratings when they give an example of what they did well or ask a question for their success criteria. Below are examples of the comments she sees:

☐ I can tell my partner my ideas about how to solve the problem.

"I figured out that we did not have to put four scooters in every box." "I'm not sure if every box has to have the same thing in it."

☐ I can find the volume of each item to be packed and use that to solve the problem.

"I found out that four scooters will fit in one shipping box."

☐ I can add volumes to find the total volume of a group of shapes.

"I explained that we could use one cube for two helmets." "I wonder why we can use one cube for two helmets."

☐ I can explain why my solution to the problem is a good one.

"The box with two scooters and 12 helmets is really heavy. I think I can figure out how to pack so none of the boxes are so heavy." "I figured out that we might ship one scooter by itself. That means there's less empty space in the boxes."

As Ms. Campbell reflects on her students' self-assessments and feedback, she is pleased with their progress. She reviews her notes from the day and thinks about her plans for the next day. She will give groups a few minutes to work together before they share their results. For groups who did not finish the task, she will ask them to identify what they have completed and what questions they still have. For groups who did complete the basic packing task, she will ask them to think about the weight of the boxes and other possible shipping strategies. She is particularly excited to learn that one group is thinking about shipping a scooter separately. She knows that group will need to share their thinking last in the rotation, as this new idea may disrupt the thinking of others. That thought pushes her to make notes about the order in which she will ask groups to share. Although she knows her plan may change during the final work time tomorrow, she has also learned that it is always more effective to plan ahead. She wants to organize the discussion around this question, "Did you pack the boxes in a different way from ___ group? Why might one way of packing be better than the other?" Therefore, she'll need to be strategic about which group presents first and how the other groups follow. This will keep creating contrasts for class discussion.

Figure 2.6 shows how Ms. Campbell made her planning visible so that she could then provide an engaging and rigorous learning experience for her learners.

Ms. Campbell's Teaching for Clarity PLANNING GUIDE

ESTABLISHING PURPOSE

1 What are the key content standards I will focus on in this lesson?

Content Standards:

> *5.MD.C. Understand concepts of volume and relate volume to multiplication and to addition.*

> *5. Relate volume to the operations of multiplication and addition and solve real-world and mathematical problems involving volume.*

> *C. Recognize volume as additive. Find volumes of solid figures composed of two non-overlapping right rectangular prisms by adding the volumes of the non-overlapping parts, applying this technique to solve real-world problems.*

Standards for Mathematical Practice:

- *Model with mathematics.*

- *Attend to precision.*

2 What are the learning intentions (the goal and *why* of learning stated in student-friendly language) I will focus on in this lesson?

- *Content: I am learning how to solve a problem about packing based on what I know about volume.*

- *Language: I am learning to describe the solution precisely using the language of volume and the language of the problem situation.*

- *Social: I am learning how to share my thinking with my partner and listen to the ideas my partner has about the problem.*

3 When will I introduce and reinforce the learning intention(s) so that students understand it, see the relevance, connect it to previous learning, and can clearly communicate it themselves?

- *Introduce learning intentions at the beginning of class to frame the task.*
- *Reinforce learning intentions throughout the class as student groups need to refocus their attention.*

SUCCESS CRITERIA

4

What evidence shows that students have mastered the learning intention(s)? What criteria will I use?

I can statements:

- *I can tell my partner my ideas about how to solve the problem.*
- *I can find the volume of each item to be packed and use that to solve the problem.*
- *I can add volumes to find the total volume of a group of shapes.*
- *I can explain why my solution to the problem is a good one.*

5

How will I check students' understanding (assess learning) during instruction and make accommodations?

Formative Assessment Strategies:

- *Observation/conference checklist with a list of anticipated strategies, success criteria, and planned questions*
- *Student work*

Differentiation Strategies:

- *Differentiate the content and product by readiness: open question*

INSTRUCTION

6 **What activities and tasks will move students forward in their learning?**

- *Packing Scooters and Helmets task*

7 **What resources (materials and sentence frames) are needed?**

Task assignment

Linking cubes for three-dimensional modeling

Rulers for creating two-dimensional models

Paper for sketching

8 **How will I organize and facilitate the learning? What questions will I ask? How will I initiate closure?**

Instructional Strategies:

- *Partner, small-group, and whole-group discussion*

Scaffolding Questions:

- *If the apostrophe were not there, what would you think about the expression?*

- *How can the diagram help you figure out the units for those values?*

Extending Questions:

- *How are you using strategies from our earlier work with volume?*

- *How are you using both multiplication and addition in this problem?*

- Can you draw a picture to show the same information as this model?

- Is this the only way the boxes can be packed? Are all of these (shipping) boxes the same weight?

Connecting Questions:

- What do you notice is the same across the representations?

- Did you pack the boxes in a different way from _____ group? Why might one way of packing be better than the other?

Self-Reflection and Self-Evaluation for Closure:

- Student self-assessment with comments

 This lesson plan is available for download at **resources.corwin.com/vlmathematics-3-5**.

Figure 2.6 Ms. Campbell's Application Lesson on Packing Scooters and Helmets

Reflection

The three examples from Ms. Buchholz, Ms. Mills, and Ms. Campbell exemplify what teaching mathematics for application of concepts and thinking skills in the Visible Learning classroom looks like. Using what you have read in this chapter, reflect on the following questions:

1. In your own words, describe what teaching for the application of concepts and thinking skills looks like in your mathematics classroom.

2. How does the planning for clarity guide support your intentionality in teaching for the application of concepts and thinking skills?

3. Compare and contrast the approaches to teaching taken by the classroom teachers featured in this chapter.

4. How did the classroom teachers featured in this chapter adjust the difficulty and/or complexity of the mathematics tasks to meet the needs of all learners?

TEACHING FOR CONCEPTUAL UNDERSTANDING

3

CHAPTER 3 SUCCESS CRITERIA:

(1) I can describe what teaching for conceptual understanding in the mathematics classroom looks like.

(2) I can apply the Teaching for Clarity Planning Guide to teaching for conceptual understanding.

(3) I can compare and contrast different approaches to teaching for conceptual understanding with teaching for application.

(4) I can give examples of how to differentiate mathematics tasks designed for conceptual understanding.

In Chapter 2, we visited the three classrooms as students engaged in the application of concepts and thinking skills. As you recall, this application of mathematics to auditorium seating, the creation of a map for the school's color run, or packaging helmets and scooters required learners to have foundational conceptual understanding and procedural knowledge. In this chapter, we will turn back time to see how each of the three teachers supported their learners as they developed conceptual understanding in their mathematics learning. We will also share videos of what conceptual learning looks like in the third through fifth grade mathematics classroom.

If learners are to see mathematics as more than algorithms and mnemonics, we must provide learning experiences that focus on the underlying properties and principles. For Ms. Buchholz, Ms. Mills, and Ms. Campbell, the end goal is for students to understand the concepts or the meaning behind mathematical procedures rather than relying on shortcuts and memory jingles. As in Chapter 2, each of these classroom teachers will differentiate the mathematics tasks by providing varying degrees of complexity and difficulty to their learners. Although every learner will be actively engaged in a challenging mathematical task that builds conceptual understanding of key concepts, Ms. Buchholz, Ms. Mills, and Ms. Campbell will adjust the complexity and difficulty of the task to ensure all learners have access to these concepts.

Ms. Buchholz and the Meaning of Multiplication

This poster hangs over the doorway of Ms. Buchholz's third grade classroom: "In this classroom: Everyone is a student. Everyone is a teacher."

Ms. Buchholz's core belief guides her instructional decisions, including her plans for this first math unit. As her students begin third grade, Ms. Buchholz knows they are embarking on a new journey as mathematicians, one where they will shift from additive thinkers to multiplicative thinkers. This is their year-long focus. She has identified four essential questions for this first unit:

- What is multiplication?
- What is division?
- How are they related?
- How do we engage in the work of mathematicians?

Although it is only the fourth day of school, Ms. Buchholz knows her students are ready to tackle the big ideas that will unite their year of mathematical work. Today, they will begin to answer "What is multiplication?" and "How do we engage in the work of mathematicians?" Ms. Buchholz has answered these questions for herself: *Multiplication is a one-to-many constant relationship between two sets that can be expressed as a ratio.* The work of mathematicians is detailed in the eight Common Core State Standards for Mathematical Practice. Today, students will begin with "make sense of problems and persevere in solving them" and "model with mathematics."

During the first few days of school, each of her students completed an interest inventory. Ms. Buchholz uses this information across the curriculum in order to connect each student meaningfully with the content and with each other. She designed today's task (The Hobbies and Activities We Love) based on students' responses.

> **EFFECT SIZE FOR TEACHER ESTIMATES OF ACHIEVEMENT = 1.29**

> **Teaching Takeaway**
>
> Effective mathematics learning involves a balance of conceptual understanding, procedural knowledge, and the application of concepts and thinking across both content and processes.

> **Teacher credibility** involves competence, trustworthiness, and caring.

> **Teaching Takeaway**
>
> Interest inventories are an effective way to build teacher-student relationships and **teacher credibility** in the classroom.

THE HOBBIES AND ACTIVITIES WE LOVE

Everyone in our class has their own favorite hobbies and activities: art, collectibles, indoor games, and outdoor games. We want to teach each other and learn about these hobbies. One option is to purchase supplies so that every

(Continued)

(Continued)

student in our class can try out every hobby. Each student would get one of each set, packet, game, or box from the categories below. There are 24 students in our class. When the order arrives, how many total pieces for each set should we receive?

Art	Indoor Games
Playdough (6 cans per set)	Board game (2 dice and 5 game pieces per game)
Scented markers (8 markers per box)	Video game (4 controllers per system)
Collectibles	**Outdoor Games**
Pokémon cards (5 or 10 cards per packet)	Kickball (4 bases and 1 ball per set)
Lego figurines (3 figurines per set)	Corn hole (8 bags and 2 boards per set)

online resources — This task is available for download at **resources.corwin.com/vlmathematics-3-5**.

Students will work with partners based on their common favorite hobby or activity. The task will engage students in beginning to answer the essential questions. The structure of the lesson will enable students to experience the routines and expectations of math class. Ms. Buchholz wants her students to see and feel her core belief in action.

What Ms. Buchholz Wants Her Students to Learn

Ms. Buchholz begins her instructional planning with the end in mind. She and her grade-level team created a profile of a third grader. This description is made up of success criteria for what an end-of-year third grader can do in the categories of content knowledge, learning and language practices across content areas, and social-emotional growth. As Ms. Buchholz plans the first math unit, she considers what structures, routines, and expectations she needs to communicate to her students in order to create a firm foundation.

EFFECT SIZE FOR TEACHER CREDIBILITY = 0.90

EFFECT SIZE FOR TEACHER-STUDENT RELATIONSHIPS = 0.52

EFFECT SIZE FOR TEACHER EXPECTATIONS = 0.43

She also relies on the Indiana Academic Standards to identify the specific elements of each unit. Ms. Buchholz identifies standards that address her essential question, "What is multiplication?"

INDIANA ACADEMIC STANDARDS

3.C.2. Represent the concept of multiplication of whole numbers with the following models: equal-sized groups, arrays, area models, and equal "jumps" on a number line. Understand the properties of 0 and 1 in multiplication.

3.AT.2. Solve real-world problems involving whole number multiplication within 100 in situations involving equal groups, arrays, and measurement quantities (e.g., by using drawings and equations with a symbol for the unknown number to represent the problem).

3.AT.4. Interpret a multiplication equation as equal groups (e.g., interpret 5×7 as the total number of objects in 5 groups of 7 objects each). Represent verbal statements of equal groups as multiplication equations.

3.AT.6. Create, extend, and give an appropriate rule for number patterns using multiplication within 100.

Ms. Buchholz is helping her learners develop the following Standards for Mathematical Practice:

- Make sense of problems and persevere in solving them.

- Model with mathematics.

> **Teaching Takeaway**
>
> Teacher clarity includes structures, routines, as well as expectations, learning intentions, and success criteria.

Learning Intentions and Success Criteria

Informed by the content and practice standards, Ms. Buchholz creates learning intentions. These are the end goals for today's lesson and stepping stones toward the profile of an end-of-year third grader:

Teaching Takeaway

Helping students take ownership of the expectations and holding themselves and their peers accountable is a characteristic of an assessment-capable visible learner.

Content Learning Intention: I am learning that the meaning of multiplication is a one-to-many constant relationship between two sets.

Language Learning Intention: I am learning the language of mathematical modeling.

Social Learning Intention: I am learning the expectations we hold for ourselves and each other as mathematicians.

In order to make these learning intentions accessible and achievable by her students, Ms. Buchholz creates the following success criteria:

☐ I can extend a one-to-many constant relationship between two sets.

☐ I can explain why the growth of each set is multiplicative.

☐ I can use mathematical models to represent multiplicative relationships.

☐ I can describe multiplicative growth using multiplicative language and notation (*equal groups, product, factor,* ×).

Ms. Buchholz knows her students will be better equipped to meet her instructional goals and evaluation criteria if she makes them transparent. Since this is their initial exploration into multiplication, Ms. Buchholz will share the essential questions while introducing the task. She does not want to give away answers to the essential questions, which are embedded in the learning intentions and success criteria. Instead, she will share the learning intentions and success criteria toward the end of their work on the task, after students have engaged in discussion about their discoveries and used these discoveries to begin to answer the essential questions in their own words.

Activating Prior Knowledge

EFFECT SIZE FOR PRIOR ABILITY = 0.94

Ms. Buchholz begins math class where the mood is filled with excitement and fun: "We're going to play a game called $10,000 Pyramid." Her students are hooked. She explains the rules and her students reorganize into pairs facing each other. The clue-giver in each pair is given a triangle of terms. The teacher explains, "All of the words are mathematical representations or tools. Get ready. Begin!" Immediately, there is a chatter of descriptions and guesses between each pair. After 2 minutes, three pairs finish and Ms. Buchholz sounds her chime. She says, "Wherever you are, stop. Reveal the words to your partner and talk through them." Ms. Buchholz hears laughter, math talk, and exclamations like, "Oh yeah, we used open number lines last year."

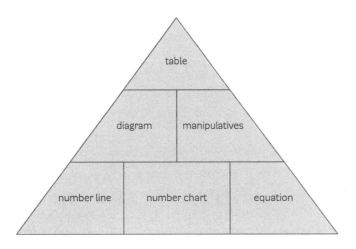

online resources | This template is available for download at **resources.corwin.com/ vlmathematics-3-5**.

"Are there any terms you want to ask questions about?"

Tahirah raises her hand and asks, "What's the difference between a picture and a diagram?"

"Who can respond to Tahirah's question?" Ms. Buchholz returns the question back to the students. They are just learning the expectation that they ask *and* answer the questions in this classroom.

"I think a diagram is a more specific kind of a picture. Like a diagram of a building is the outline of the building with the rooms labeled," responds Frances.

"I made a diagram of a butterfly life cycle last year. I sketched each part of the life cycle and then wrote labels. I also drew lines to label the body parts of a caterpillar and butterfly," adds Chase.

Ms. Buchholz summarizes, "So a diagram is a picture but it has labels to name the parts. It may just be an outline or a sketch. Tahirah, does that answer your question?" Tahirah nods.

"This is your mathematical toolbox," Ms. Buchholz says as she passes out a chart to each student and offers the following explanation (Figure 3.1).

> One of our goals this year is to gather mathematical tools, learn how to use them, and then add them to our toolbox so we can use them whenever we need. Right now, your toolbox has the six tools from $10,000 Pyramid. We will learn more about each of them and, gradually, we'll add more tools.
>
> We're going to take a moment so you and your partner can make some notes. This is *your* toolbox, so you write what you know right now. We will add more to it later.

EFFECT SIZE FOR NOTE TAKING = 0.50

Ms. Buchholz gives students 3 minutes to talk, write, and sketch.

She then says, "Today, we start our first unit in math. These are our four essential questions that we will work toward answering over the next few weeks. Read the questions and think about what they mean." Ms. Buchholz points to where the unit essential questions are displayed and allows silence while everyone reads and thinks.

She continues, "Now pair up and talk about these four essential questions. Share what you are thinking about with your partner." After a minute of talk, she brings everyone back to together and says, "Let's have two people share what they said or what their partner said."

EFFECT SIZE FOR PLANNING AND PREDICTION = 0.76

Ty begins, "My sister showed me some multiplication, like two times two is four."

Dillon adds, "Sabina and I were thinking our mathematical toolbox might help answer the question about mathematicians' work, if mathematicians actually use those tools."

Ms. Buchholz says, "Today, we are going to focus on just two of the essential questions: *What is multiplication? How do we engage in the work of mathematicians?* We're going to start our work with a task based on our interest inventory." She passes out The Hobbies and Activities We Love task to each student and says, "I'm going to read the task and you

THE MATHEMATICAL TOOLBOX

_____'s Mathematical Toolbox

Tool	Manipulatives	Diagram	Number Line	Number Chart	Table	Equation
Description						
Example						
When I Use It						

Source: Toolbox Image copyright Lazarev/iStock.com

online resources ➤ This template is available for download at **resources.corwin.com/vlmathematics-3-5**.

Figure 3.1

can read along with me." She knows that not all of her students are able to read this text independently. As Ms. Buchholz reads aloud, she pauses and models thinking aloud at critical points she planned in advance.

> Every student is going to get one set. So we need 24 playdough sets, 24 Lego figurine sets, and 24 game boards. You will work with a partner who has a similar favorite hobby or activity.

> Together, you will be responsible for finding the total pieces of each set in your category. For example, my favorite activity is indoor games. So I would be finding the total pieces for 24 board games and the total controllers for 24 video games. Find your partner and sit with them.

Ms. Buchholz displays a list of pairs and favorite hobbies or activities. Her students find their partners and sit down again. She wants to make sure students comprehend the problem before she lets them get to work. She uses guided questioning to help her students apply literacy comprehension strategies to this task (Thunder & Demchak, 2012).

EFFECT SIZE FOR COOPERATIVE VS. INDIVIDUALISTIC LEARNING = **0.55**

> Close your eyes and visualize the art supplies. Can you see the boxes of playdough? Each box has six cans of playdough in it. Can you see the scented markers with eight markers in each box? We need 24 boxes of playdough and 24 boxes of scented markers.

> Now, open your eyes. I'm looking at this list and making a connection to when I was buying school supplies. I bought five boxes of pencils and there were 10 pencils in each box. Who else has a connection to this problem?

Ms. Buchholz calls on two additional students to share their connections before she continues.

> I would predict or estimate that our order would have more than 24 of each piece because we need 24 of each set and each set has more than one piece. I estimate our order would have less than 1,000 of each piece because that's a really big number and I think we might order a lot. Who agrees or disagrees with my estimates?

As she pauses, Ms. Buchholz then calls on a student to explain her reasoning. Akeiyla disagrees: "I think 1,000 is too much. Maybe more like 200 would be the most because six and six is 12 so that's two sets. And we need 24 sets but that won't be as much as 1,000."

"I have a clarifying question," Ms. Buchholz says. "What's the difference between a set and total pieces?" As students take turns responding, Ms. Buchholz also prompts them to ask their own clarifying questions of the group. Some of their comments are as follows: "Is a packet a set? Is a game the same thing as a set?" "For some hobbies, it shows two sets, like kickball. Are we figuring out how many bases and how many balls?" "For Pokémon cards, it says 'or'. Five *or* 10 cards per packet. Do we have to figure out both? Can we just choose one?" Ms. Buchholz has students ask and answer questions. For some, they come to a class consensus, such as "Yes, you figure out both sets: how many bases and how many balls" and "Even though it says 'or,' you still have to figure out both five cards per packet and 10 cards per packet because we don't know for sure which one we'd order."

Ms. Buchholz can tell from her students' questions and responses that they are each gaining entry into the task in meaningful ways. Her last literacy comprehension strategy is summarization: "Who can summarize what we're trying to figure out?" Several students volunteer.

> EFFECT SIZE FOR
> SUMMARIZATION
> = **0.79**

The students' prior knowledge is activated and the task is understood. Ms. Buchholz's final step with the whole class is to make expectations clear for what this work time will look like and sound like as well as what students should be ready to share after work time (Van de Walle et al., 2018).

As you problem solve, you should collaborate with your partner. That means you talk, share materials, and make decisions together. You should choose tools from your mathematical toolbox for problem solving. You should share ideas and come to consensus while also recording your thinking and models on your own paper. I will confer with partners as they work. When we come back together, I will ask some pairs to share the tools from the mathematical toolbox that worked or didn't work for solving this problem.

> EFFECT SIZE FOR
> STRATEGY TO
> INTEGRATE WITH
> PRIOR KNOWLEDGE
> = **0.93**

Ms. Buchholz allows a few moments for final questions and then lets the pairs go work.

EFFECT SIZE FOR
SELF-EFFICACY
= 0.92

EFFECT SIZE FOR
RECORD KEEPING
= 0.52

Scaffolding, Extending, and Assessing Student Thinking

Each pair finds a space to work and confirms which category they are responsible for solving. Some students get additional materials, while others talk, write, and draw. The room is not quiet, but every pair is on task.

Ms. Buchholz knows it's important for all students to see themselves as capable problem-solvers and mathematicians. Conferences serve as an opportunity for her to scaffold, extend, and formatively assess student thinking. By documenting her conferences and observations, Ms. Buchholz is able to use her notes and student work to plan the next instructional steps and differentiate future tasks.

For today's task, Ms. Buchholz creates a conference/observation chart with two sections (Figure 3.2). The first is focused on the third success criterion: *I can use mathematical models to represent multiplicative relationships*. The six models from the mathematical toolbox are listed so she can note who is using which representation as well as who will share and in what sequence (Fennell, Kobett, & Wray, 2017). The second section is focused on the other three success criteria. There is open space to note how each student is demonstrating the success criteria or misconceptions. Ms. Buchholz also includes the teacher questions she has planned for scaffolding and extending student thinking for quick reference.

Adam and Kayvion have a page filled with numbers. Ms. Buchholz joins them and asks, "What are you working on?"

"We're figuring out the number of bases in 24 sets of kickball games. This is our list so far: 4, 8, 12, 16, 20, 24, 28," Adam points to a scattered list of numbers.

"What are these numbers?" Ms. Buchholz points to another row of numbers connected to each of the numbers Adam has described by a line.

"Those are how many balls: 1, 2, 3, 4, 5, 6, 7. That's easy because it's just one more ball each time," Adam explains.

Kayvion adds, "And these numbers up here are how many sets we've counted so far so we know when we get to 24 total," pointing to yet another set of numbers connected by more lines.

The Hobbies and Activities We Love Task

Observation/Conference Chart

Date _____

Questions:

- What tool from your toolbox could you use?
- How is _____ changing each time?
- What pattern do you notice?
- How many _____ would you need for 10 students or 20 students? How do you know?
- How are the quantities changing?

Name	SC 3: I can use mathematical models to represent multiplicative relationships. Concrete: manipulatives Representational: diagram, number line Abstract: number chart, table, equation	SC 1: I can extend a one-to-many constant relationship between two sets. SC 2: I can explain why the growth of each set is multiplicative. SC 4: I can describe multiplicative growth using multiplicative language and notation (*equal groups, product, factor, ×*)

Figure 3.2

Ms. Buchholz shares her noticing of their strategy and says, "I see you keeping track of three kinds of numbers: how many sets, how many balls, and how many bases. Those are really important for solving this problem and they're all related." Then she interprets their understanding and decides how to respond to move them forward (Jacobs, Lamb, & Philipp, 2010). Ms. Buchholz poses a question to assess her students' prior knowledge about the tools: "You also need to show that relationship. Right now, you're using lines to connect the numbers. But your numbers look a bit scattered and may be hard to keep track of as you reach 24 sets. Let's look at your mathematical toolbox. Which tool might help you to organize this list of numbers?"

"We're using a number chart to help us count four more bases for each set. But it's not helping us keep track of the number of balls or sets," Kayvion reflects.

"Our list of numbers looks kind of like a number line. We could try that," Adam says.

"I remember using a table last year to keep track of numbers too," Kayvion adds.

"Which would you like to try—a number line or a table?" asks Ms. Buchholz. Both students choose the table. Ms. Buchholz decides to teach a mini-lesson on setting up a table for Adam and Kayvion and says, "A table has categories and then numbers. What are your categories?" She sketches a table on a separate piece of paper based on Adam and Kayvion's responses. They begin adding the numbers to the table under each category and extending the patterns while referring to the number chart. Ms. Buchholz pauses to add notes to her chart.

Before conferring with another pair, Ms. Buchholz observes the class as a whole. She notes that Kadisha and Frances are stapling paper to make long number lines. Chase and Briera seem stuck.

"What are you two working on?" Ms. Buchholz asks as she sits down with Chase and Briera.

Chase responds quietly, "I don't know. Collectibles."

"We started with the Lego figurines but we got stuck," Briera adds.

Ms. Buchholz researches to find out more about their thinking. "What have you tried so far?" she asks.

Chase points to their paper and says, "We drew all these circles and wrote three in them but now we don't know what to do."

"Why did you write three inside each circle?" the teacher asks.

"Because there are three figurines in each set," he responds.

"And how many circles did you draw?" Ms. Buchholz questions.

"Twenty-four because we need to order 24 sets for the whole class to get one each," Briera says.

"What are you trying to figure out now?" Ms. Buchholz asks.

"How many figurines there are. But that's a lot of threes to add up," Chase replies.

Ms. Buchholz summarizes their reasoning by saying, "You've figured out how many figurines in each set and how many sets you need all together. So now you're adding up all of these threes. Let's look at your mathematical toolbox to see what tool might help you." She then refers the students to the scaffold of the graphic organizer.

"Can we use cubes?" they ask.

"Cubes are a type of manipulative," Ms. Buchholz says as she points to manipulatives on the graphic organizer. "How might you use cubes to add up the threes?"

"We could put three cubes on each circle and then snap them together," says Chase.

"Yeah, we could snap them into groups of tens and ones and then count them," says Briera.

"How many piles of three cubes will you make?" asks Ms. Buchholz.

"Twenty-four," Chase replies.

Ms. Buchholz describes their problem-solving process and offers a solution to support their metacognition and agency: "That sounds like a plan. You are using two tools together to help you problem solve. You have the start of a diagram with your circles and you'll use manipulatives. Using more than one representation can help when you get stuck." Then she makes notes in her chart, including a note to have them connect their manipulatives and picture to an equation. Ms. Buchholz knows that using a sequence of concrete-representational-abstract models for problem solving and making connections among the models is an effective scaffold for all students (Berry & Thunder, 2017).

Again, Ms. Buchholz surveys her class and makes observation notes. Nathan asks if he can use a calculator even though it's not listed on the mathematical toolbox. Ms. Buchholz asks him how a calculator would help him and then shows him where they are. She makes a note about this interaction and joins another pair for a conference.

By the end of work time, Ms. Buchholz has observational or conference notes on every student. She has also selected and sequenced four student pairs who will share based on the representations they used (Smith & Stein, 2011):

1. Chase and Briera will share their manipulatives and diagram.

2. Adam and Kayvion will share their table and number chart.

3. Frances and Kadisha will share their number line.

4. Nathan and Geff will share their use of the calculator.

Teaching for Clarity at the Close

As each of the four pairs shares their representations with the whole class, Ms. Buchholz asks them to write an equation to represent their work. The students write repeated addition equations, such as $3 + 3 + 3 + 3 + 3 + 3 + 3 + 3 + 3 + 3 + 3 + 3 + 3 + 3 + 3 + 3 + 3 + 3 + 3 + 3 + 3 + 3 + 3 + 3 = 72$. She also asks the following focusing questions to highlight the connections among the representations, the usefulness of the representations, and the multiplicative relationships:

- Where do you see the number of students represented?
- Where do you see the number of _____ represented?
- How are the quantities changing?

- How does this model represent the relationship between the two quantities?

- How does this model help you clearly communicate your strategy and thinking?

- How does this model help you efficiently solve the problem?

- What is similar among the models? What is different?

Her questions align with the essential questions and learning intentions to develop students' conceptual understanding of multiplication.

Ms. Buchholz notices her students' bringing their prior knowledge of skip counting, addition, and patterns to bear on this new operation. They are engaged in surface-level learning related to multiplication.

> I hear you describing a constant relationship between one set and the many pieces within the set. As you increase the number of sets, the number of pieces also grows at a constant rate. Many of you were skip counting or adding the same amount over and over to extend the pattern. Talk with your partner about how you would answer each of the essential questions for today based on our work.

The students are ready to connect their informal language to the formal school language of multiplicative thinking. Ms. Buchholz uses this opportunity to explicitly introduce the academic language and notation of multiplication in a meaningful way. She says, "Mathematicians like to be efficient in their problem solving and their notation. Many of you commented on how long your equations were and how easy it is to skip a number. So mathematicians use a more efficient and accurate notation." She returns to one repeated addition equation and writes it in informal words saying, "I can also read this equation as 24 equal groups of 3 is the same as 72 and the mathematical notation would be . . ." She writes the multiplication equation, $24 \times 3 = 72$, and repeats the informal language. Then, she labels the factors and product. She says, "With your partner, return to your equation and write it in words and then write it in this efficient notation." Each pair works for a moment and then shares their new equations.

EFFECT SIZE
FOR DIRECT/
DELIBERATE
INSTRUCTION
= 0.60

Then Ms. Buchholz shares the learning intentions and success criteria for today's work. She says, "Each day, we will have learning intentions or goals for the day and success criteria or ways to evaluate your work to know if you met the goals completely or partially. Today, we will evaluate our work as a class together." Ms. Buchholz reads each learning intention and its related success criteria. Then she asks pairs to combine with another pair and discuss whether they completely or partially met the success criteria today. She says, "Also discuss evidence you can cite to defend your evaluation." This is the first time this year that her students are using learning intentions and success criteria to orient their work. Ms. Buchholz knows she has to teach her students what these are and how they can guide their work.

After discussing each success criterion in their small group, the whole class shares their ideas. Ms. Buchholz models writing their evidence on a green (got it completely), yellow (partially got it), or red sticky note (stuck or confused) and placing the sticky notes next to each success criterion. Then she summarizes their progress to initiate closure.

> EFFECT SIZE FOR
> META-COGNITIVE
> STRATEGIES = 0.60

As a class, we can extend a one-to-many constant relationship between two sets: got it! We can explain why the growth of each set is multiplicative: we're confused. We can use mathematical models to represent multiplicative relationships: got it! And we are starting to describe multiplicative growth using multiplicative language and notation.

Finally, she has students place their work and mathematical toolboxes in their math binders. Tomorrow, they will begin by adding a new tool to their mathematical toolbox graphic organizer (the calculator), as well as writing descriptions, examples, and evaluations of when each tool is useful. Ms. Buchholz knows her students need more experiences to understand why the growth of sets is multiplicative, and this is confirmed by the class evaluation.

Figure 3.3 shows how Ms. Buchholz made her planning visible so that she could then provide an engaging and rigorous learning experience for her learners

Ms. Buchholz's Teaching for Clarity PLANNING GUIDE

ESTABLISHING PURPOSE

1

What are the key content standards I will focus on in this lesson?

Indiana Academic Standards:

> 3.C.2. Represent the concept of multiplication of whole numbers with the following models: equal-sized groups, arrays, area models, and equal "jumps" on a number line. Understand the properties of 0 and 1 in multiplication.

> 3.AT.2. Solve real-world problems involving whole number multiplication within 100 in situations involving equal groups, arrays, and measurement quantities (e.g., by using drawings and equations with a symbol for the unknown number to represent the problem).

> 3.AT.4. Interpret a multiplication equation as equal groups (e.g., interpret 5×7 as the total number of objects in 5 groups of 7 objects each). Represent verbal statements of equal groups as multiplication equations.

> 3.AT.6. Create, extend, and give an appropriate rule for number patterns using multiplication within 100.

Standards for Mathematical Practice:

- Make sense of problems and persevere in solving them.

- Model with mathematics.

2

What are the learning intentions (the goal and *why* of learning stated in student-friendly language) I will focus on in this lesson?

- Content: I am learning that the meaning of multiplication is a one-to-many constant relationship between two sets.

- Language: I am learning the language of mathematical modeling.

- Social: I am learning the expectations we hold for ourselves and each other as mathematicians.

3 When will I introduce and reinforce the learning intention(s) so that students understand it, see the relevance, connect it to previous learning, and can clearly communicate it themselves?

- Introduce and discuss related essential questions first: What is multiplication? How do we engage in the work of mathematicians?

- Facilitate partner and class discussion to answer essential questions.

- Facilitate small group and class discussion of learning intentions and success criteria with modeling of sticky notes and evidence to close the lesson.

SUCCESS CRITERIA

4 What evidence shows that students have mastered the learning intention(s)? What criteria will I use?

I can statements:

- I can extend a one-to-many constant relationship between two sets.

- I can explain why the growth of each set is multiplicative.

- I can use mathematical models to represent multiplicative relationships.

- I can describe multiplicative growth using multiplicative language and notation (equal groups, product, factor, ×)

5 How will I check students' understanding (assess learning) during instruction and make accommodations?

Formative Assessment Strategies:

- Observation/conference chart

- Student work

- Colored sticky notes with evidence

Differentiation Strategies:

- Differentiate the content and process by interest: partner and task category based on interest inventory

- Differentiate the process by interest: choice of tools from the mathematical toolbox

INSTRUCTION

6 What activities and tasks will move students forward in their learning?

- $10,000 Pyramid

- The Hobbies and Activities We Love task with the mathematical toolbox (concrete, representational, and abstract [CRA] representations)

7 What resources (materials and sentence frames) are needed?

$10,000 Pyramid terms

Mathematical toolboxes

Math binders

Cubes

Number charts

Graph paper

Open number lines and whiteboard markers

Colored pencils

Calculators

8 How will I organize and facilitate the learning? What questions will I ask? How will I initiate closure?

Instructional Strategies:

- Literacy comprehension strategies: visualizing, making connections, predicting, asking questions, and summarizing

- Anticipate, monitor, select, sequence, and connect students' CRA strategies
- Think-pair-share

Scaffolding Questions:

- What tool from your toolbox could you use?
- How is _____ changing each time?

Extending Questions:

- What pattern do you notice?
- How many _____ would you need for 10 students or 20 students? How do you know?
- How are the quantities changing?

Connecting Questions:

- Where do you see the number of students represented?
- Where do you see the number of _____ represented?
- How are the quantities changing?
- How does this model represent the relationship between the two quantities?
- How does this model help you clearly communicate your strategy and thinking?
- How does this model help you efficiently solve the problem?
- What is similar among the models? What is different?

Self-Reflection and Self-Evaluation for Closure:

- Colored sticky note class evaluation

 This lesson plan is available for download at *resources.corwin.com/vlmathematics-3-5*.

Figure 3.3 Ms. Buchholz's Conceptual Understanding Lesson on the Meaning of Multiplication

Ms. Mills and Representing Division as Fractions

Ms. Mills has arrived at her favorite time of year. Her fourth grade students are multiplicative thinkers who are ready to delve into rational numbers. Rational numbers and proportional reasoning will be the foundation of her students' future mathematical learning through college. The rational numbers unit revolves around the big idea that fractions are another way to represent division. Any fraction's numerator is a dividend and that same fraction's denominator is the divisor. In other words, $\frac{1}{4}$ is really 1 divided by 4.

Ms. Mills knows that her students have experiences with fractions from previous grades as well as their daily life. They bring this rich experiential knowledge along with their more recently developed multiplicative thinking on the meanings of multiplication and division, representations and mathematical tools for multiplication and division (such as ratio tables and double number lines), and fluency with multiples and factors.

During science, the class has been studying the conservation status of animals around the world, animal populations, and the reasons behind their endangerment. Ms. Mills has created contextualized problems that will garner their interest by extending this study into mathematics. She presents students with an anchor problem that all students will complete as well as several extension problems from which students will choose one or more to complete (Thunder, 2014). The anchor problem is an equalizing problem (Empson & Levi, 2011) where students will use what they know about equal sharing from multiplication and division and transfer this to fractions.

> ☐ *Anchor Problem:* At one zoo, four giant pandas eat 1 ton of bamboo each week. At another zoo, there are eight giant pandas. How many tons of bamboo per week should they get so that each giant panda gets the same size share of bamboo as a giant panda at the first zoo? How much of a ton will each giant panda eat in a week?
>
> *(Continued)*

Teaching Takeaway

Sharing the big ideas helps learners see where they are going in their learning.

> **EFFECT SIZE FOR PRIOR ABILITY = 0.94**

> **EFFECT SIZE FOR INTEGRATED CURRICULA PROGRAMS = 0.47**

(Continued)

Two of the extension problems are as follows:

☐ *Extension Problem 1:* Two white rhinos in a zoo can consume 7 tons of grasses per month. In the wild, white rhinos live in groups called crashes. A crash can be made of up to 14 white rhinos. Assuming each white rhino gets the same size portion of grasses in the wild as in a zoo, how many tons of grasses would a crash of white rhinos need access to each month?

☐ *Extension Problem 2:* Three tigers at a zoo typically eat 1 ton of meat during the summer months. Zoologists studying the 17 tigers in their natural habitat of Laos are concerned that the tigers do not have enough prey to survive. How many tons of meat do the Laos tigers need to consume to get the same size share of meat as tigers thriving in a zoo?

What Ms. Mills Wants Her Students to Learn

Ms. Mills breaks broader learning intentions down into meaningful chunks that build upon each other and move her students from surface learning to deep learning to transfer learning. She begins her rational numbers unit focused on the overarching big idea of the unit: *Fractions are another way to represent division.* This will require students to transfer their understanding and knowledge of division to new situations where there are remainders to be shared equally. Her first lesson of the unit will focus on the following standard.

EFFECT SIZE FOR
COGNITIVE TASK
ANALYSIS = **1.29**

VIRGINIA MATHEMATICS STANDARDS OF LEARNING

4.2c. The student will identify the division statement that represents a fraction, with models and in context.

> **Ms. Mills is helping her learners develop the following Mathematical Process Standards:**
>
> - Mathematical communication
>
> - Mathematical problem solving
>
> - Mathematical representations

Ms. Mills unpacks this standard carefully to make sure she addresses each component. She also keeps the end in mind: fourth graders will transfer their knowledge again toward the end of this unit to connect fractions with decimals.

Learning Intentions and Success Criteria

Ms. Mills knows that teacher clarity is important to her students' success. She creates three learning intentions that encompass the content and process standards:

> *Content Learning Intention*: I am learning to understand the connections among representations of fractions.
>
> *Language Learning Intention*: I am learning to understand the language that makes meaning of fraction notation.
>
> *Social Learning Intention*: I am learning to understand how to listen and respond to my peers' ideas in ways that move us all forward as learners.

Based on these learning intentions, Ms. Mills creates success criteria so that her students can see themselves learning and evaluate their movement along the path to success. Her success criteria communicate to students what they will be able to do and why this matters:

Teaching Takeaway

Balanced mathematics instruction requires us to have clarity about what our learners must know, understand, and be able to do.

EFFECT SIZE FOR ASSESSMENT-CAPABLE VISIBLE LEARNERS = **1.33**

☐ I can represent fractions using multiple representations.

☐ I can explain the connections among these representations.

☐ I can use these representations to prove fractions are equivalent.

☐ I can communicate my process, my representations, and my mistakes using fraction language: *number of pieces (numerator)*, *size of piece (denominator)*, and *same size share (equivalent fractions)*.

Video 7
Choosing a Conceptual
Learning Task

*https://resources.corwin.com/
vlmathematics-3-5*

EFFECT SIZE FOR
VOCABULARY
INSTRUCTION
= 0.62

Ms. Mills uses mathematically accurate and important language in her learning intentions and success criteria while also making them concise and child friendly. She knows her fourth graders are always learning new vocabulary. In order for this language to become part of their receptive and expressive vocabulary, her students need multiple and various exposures to mathematical language and experiences that connect new words to nonlinguistic representations (National Reading Panel, 2000). The learning intentions and success criteria are one opportunity to make this vocabulary instruction explicit. The task, discourse, and teacher questioning within the lesson are additional opportunities.

During her lesson, Ms. Mills plans to share the learning intentions and success criteria with her students after activating their prior knowledge. As students are engaged in surface learning, she will reconnect them with familiar fraction language, notation, and representations through problem solving, think-alouds, and purposeful questioning.

Activating Prior Knowledge

Around the room, there are five posters hung on the walls. Each one has a large word and a smaller question beneath it:

- *Context:* When would you need to use a fractional amount?

Context
When would you need
to use a fractional amount?

- *Language:* What words do you use to describe fractional amounts?

> Language
> What words do you use
> to describe fractional amounts?

- *Common Mistakes:* What types of mistakes do people make when working with fractions?

> Common Mistakes
> What types of mistakes do
> people make when working
> With fractions?

- *Representations:* How can you represent fractional amounts—in school and in real life?

> Representations
> How can you represent fractional
> amounts -in school and in real life?

- *Tools:* What tools measure fractional amounts?

> Tools
> What tools measure
> fractional amounts?

Ms. Mills uses this Graffiti activity to activate her students' prior knowledge related to the entire rational numbers unit. Each student is given a marker. They may draw, list, label, write sentences, or choose another approach as long as they do not write over anyone else's ideas and keep the marker to the paper the whole time. Four or five students stand at each poster. They rotate every minute and finally return to their starting poster. As a team, they read the poster in order to develop a one- to two-sentence answer to the question. After 3 minutes of reading and talking time, each group shares their word, question, and answer.

Next, Ms. Mills transitions the class to the task at hand by introducing today's learning intentions and success criteria. Her students are in the

> EFFECT SIZE FOR STRATEGY TO INTEGRATE WITH PRIOR KNOWLEDGE = 0.93

> EFFECT SIZE FOR COOPERATIVE LEARNING = 0.40

habit of reading the learning intentions and success criteria and explaining to a partner what they think each means, describing when they have done something similar, and asking each other clarifying questions. Then, Ms. Mills shares the anchor problem.

> We've been studying animals at different levels of conservation status from extinct to endangered to threatened. We know that a lack of food resources is one reason some animals are endangered. We also know the quantity of food resources required to keep an endangered animal at a zoo impacts conservation. Today, we are going to work on an anchor problem related to the food consumption of giant pandas. Then you can work by yourself or with a partner to choose which extension problems you would like to solve.

As Ms. Mills speaks, she passes out the anchor problem and students glue it into their mathematician's journals—legal-size paper stapled together with a fold 2 inches from the end. In the large part of each journal page, students glue their anchor and extension problems and record discoveries, lists, pictures, charts, and diagrams that support their problem-solving process and evaluation of their solution. On the folded part, students record questions and answers, connections, reflections, and big ideas. Together, the two sections of the mathematician's journal require students to keep a record of their process and solution while also self-monitoring. Students also organize and transform their work as a reference for sharing and future problem solving.

Ms. Mills reminds students of the STAR (Search, Translate, Answer, and Review) strategy (Gagnon & Maccini, 2001). "With your elbow partner, read over the anchor problem. Use the first step in the STAR strategy: *Search* for important information." Pairs of students read and talk. Some use highlighters. Others circle, underline, and write phrases and questions with colored pencils. Ms. Mills listens in on students' conversations and then asks them to share their *Search* findings.

"There are two zoos. One zoo has four giant pandas and the other has eight giant pandas."

"The first zoo has four pandas. They eat 1 ton of bamboo."

"In a week! They must eat a lot in 1 day!"

"How would a zoo grow enough bamboo to feed them?"

EFFECT SIZE
FOR CLASSROOM
DISCUSSION = 0.82

The students talk to each other. As one student stops speaking, he or she acknowledges the next person to speak. Then Ms. Mills asks, "So what are we trying to figure out?"

"How much to feed the eight pandas."

"Every panda has to get the same amount, so how much do they each get?"

"But how many tons of bamboo do we have?"

"Yeah I don't know. It says the first zoo has 1 ton. But I don't know how much the second zoo has."

"That's what we're trying to figure out. We get 1 ton for four pandas. So how much should we get for eight pandas?"

"Oh right! Every panda at the first zoo gets an equal share of 1 ton. And every panda at the second zoo should get the same size share."

The teacher allows the students to talk, pose questions, answer each other's questions, express confusion, and seek clarity. She knows spending this time to make sure the problem is understood will pay off as students move on to selecting representations and strategies and to checking the reasonableness of solutions.

EFFECT SIZE FOR
HELP SEEKING
= 0.72

Ms. Mills says, "The next step of STAR is *Translate*. Translate the words into models, pictures, or symbols." She refers to an anchor chart that lists the four steps of the STAR strategy for problem solving:

> *Search* the word problem for important information (not key words). *Translate* the words into models, pictures, or symbols. *Answer* the problem. *Review* your solution for reasonableness.

> We've used a lot of new models in our multiplication and division unit, like ratio tables and double number lines. Those may come in handy. We also know models that are tried and true for us, like bar models and a sketch. Choose a model that you might translate the words of this problem into and

Teaching Takeaway

We can use purposeful questioning to both guide and focus learners' thinking.

Video 8
Making Thinking
Visible and
Addressing Roadblocks

*https://resources.corwin.com/
vlmathematics-3-5*

visualize how you would make the model. How will you represent the first zoo with four pandas and 1 ton of bamboo? How will you represent the second zoo with eight pandas and some unknown amount of bamboo? How will you represent the pandas sharing the bamboo equally? Turn and tell your elbow partner what you are visualizing.

Again Ms. Mills listens in on conversations and asks focusing questions.

I heard a lot of great visualizing. First, I'm going to share my image. I can see one option is sketching a bar model. The first zoo could look like this: one long bar for 1 ton of bamboo and four equal smaller bars for the four pandas sharing that bamboo. Then the second zoo would also have a long bar—that mystery bar that we're wondering about—and eight equal smaller bars for the eight pandas sharing that bamboo. I also heard Charles describing a double number line. Charles, what did you visualize?

Charles says, "I saw two open number lines on top of each other like this." He draws two parallel, horizontal lines in the air. "And then two vertical lines like this." He draws two parallel, vertical lines in the air. As Charles talks, Ms. Mills draws a sketch of the double number line. "The top number line is 'Tons of Bamboo' and the bottom is 'Pandas.' I'd write 1 and 4 and then? And 8." Ms. Mills labels the double number line and continues.

We *searched* the problem for important information and visualized some ways to *translate* the words into models, pictures, or symbols. Next, you will *answer* the problem and then *review* your solution for reasonableness. There are lots of materials available. You can work alone or with a partner as you complete the anchor problem and move on to the extension problems. Remember to keep track of your strategies, your solution, and your thinking in your mathematician's journal. I will be conferring with you as you work. During our last 15 minutes, we will share our processes and reflect back on our learning intentions and success criteria.

Scaffolding, Extending, and Assessing Student Thinking

The classroom swiftly becomes a buzz of activity as students rearrange to find suitable work spaces, desired materials, and partners of interest. Ms. Mills confers with individuals and pairs while taking notes on her conference notes chart. When she meets with students who are struggling, she connects them with a manipulative and asks focusing questions such as "How could these cubes represent what we know? How will you know when each panda has an equal share?" There are some students who easily see the fourths and doubling relationship in the anchor problem. They move on to the extension problems, and Ms. Mills asks them higher-level questions such as "What if there were five pandas in the first zoo? How would this change the solution? What if there were nine pandas in the second zoo?"

Ms. Mills also uses her conference notes to document students' progress toward the learning intentions by listening and looking for evidence of the success criteria. Today, Ms. Mills uses an open conference chart, where each student's name is listed with space for notes. Her students are engaged in surface learning and she knows her students will surprise her with what they know, notice, ask, and try. The openness of her conference chart allows Ms. Mills to document both what she has anticipated and what she has not.

As she confers with students, Ms. Mills looks for three strategies to select for sharing (Smith & Stein, 2011): someone who used a concrete representation of division, a student who used a pictorial representation, and another student who used an abstract representation. The learning intentions and success criteria are listed at the top of her chart. Her planned scaffolding and extending questions are listed at the bottom of her chart for quick reference and to keep her notes and questioning focused on the goals of the lesson.

Teaching for Clarity at the Close

Ms. Mills dings her chime. The class reorganizes themselves on the carpet, with their mathematician's journals and colored pencils in hand. The teacher says, "We worked hard as mathematicians today. We are

EFFECT SIZE FOR CONCENTRATION/ PERSISTENCE/ ENGAGEMENT = 0.56

Teaching Takeaway

Conferring with learners is an effective way to formatively evaluate their learning progress.

EFFECT SIZE FOR SCAFFOLDING = 0.82

EFFECT SIZE FOR TEACHER EXPECTATIONS = 0.43

Teaching Takeaway

Visible Learning in the mathematics classroom requires that we have clarity about what learners are expected to know, understand, and be able to do in their learning.

**Teaching
Takeaway**

Visible Learning in
the mathematics
classroom involves
the gathering of
evidence about
the impact we
have on our
students' learning.

working to examine the connections among representations of fractions and to describe these connections using the language and notation of fractions. As we share, remember to listen to and respond to each other's ideas and questions in ways that move us all forward as learners." Ms. Mills gestures to the dialogue bubbles with language frames around her board (Figure 3.4).

Her students have internalized some of these frames and use them throughout sharing conversations, while others remain as a reference.

Ms. Mills sequences her sharers from concrete to representational to abstract (Berry & Thunder, 2017). First, Lucas shares his concrete representation of division. Lucas used Cuisenaire rods and then labeled a sketch of the Cuisenaire rods in his mathematician's journal. Next, Leah shares her representational strategy. She drew a bar model.

Students make connections between the representations. Kamora says, "I was thinking about Lucas's rods and I noticed that Leah's bar model looks just like his rods except without the color."

LANGUAGE FRAMES

- I agree with _____ because . . .
- I disagree with _____ because . . .
- I still have questions about. . . .
- To clarify, you're saying. . . .
- I was thinking about what _____ said, and I was wondering, what if . . .
- I don't know what you mean by . . .
- I want to go back to what _____ said . . .
- I want to add on to what _____ said about . . .

Figure 3.4

"I agree with Kamora because Lucas had red rods that were one-fourth and brown rods that were 1 ton. Leah also has fourths and one whole," Calvin adds.

Ms. Mills says, "Turn to a partner and explain the connection between Lucas's and Leah's representations. How are they similar? How are they different?" After 1 minute of discussion, Ms. Mills continues, "So far we have seen two representations for fractional amounts—Cuisenaire rods and a bar model. You are noticing and explaining powerful connections among these representations, which are two of our success criteria. Keep looking for these connections as Zahara and Saw share their strategy."

Zahara and Saw share their abstract representation (a ratio table) as well as a mistake they made and how they recognized the mistake and corrected it.

"Let's look at all three representations," Ms. Mills says. She poses many of her planned connecting questions to facilitate the discussion:

- Where do you see one-fourth in each?

- Where do you see two-eighths in each?

- How is doubling related to equivalence in this problem?

- What representations were efficient, flexible, and accurate? What representations were not? Why?

- How is 1 ton divided by four pandas related to $\frac{1}{4}$? And 2 tons divided by eight pandas related to $\frac{2}{8}$?

She uses these focusing questions to help students make connections across the representations that will help them understand the lesson's big idea: *Fractions are another way to represent division.*

After more discussion, Ms. Mills initiates closure. She knows that this is the first day of the unit. Most of her students are at the surface level of learning, while some are moving into the deep level of learning. Many questions remain. Not everything is clear and comfortable. Ms. Mills knows this cognitive dissonance is important for

Teaching Takeaway

As part of the closure to a lesson, linking back to the learning intentions and success criteria enhances the clarity in the learning for students and supports the building of assessment-capable learners.

> EFFECT SIZE FOR TEACHING COMMUNICATION SKILLS AND STRATEGIES = 0.43

> EFFECT SIZE FOR EVALUATION AND REFLECTION = 0.75

> EFFECT SIZE FOR STRONG CLASSROOM COHESION = 0.44

Teaching Takeaway

To support learning, we must offer students opportunities to engage in dialogue about mathematics content and processes.

students' learning and she is willing to stand in the murkiness with them. However, she also wants to make the day's learning visible to the students themselves.

> We're going to close today's work with some personal reflection time. This time is for you as a mathematician and learner. First, what do you want to record on the flap of your mathematician's journal? What aha moment have you had? What mistake do you want to keep track of? What will you take away from today's work? Record your thoughts.

Students use their colored pencils to record their self-monitoring reflections.

> Now let's return to our learning intentions and success criteria. Remember, the success criteria let you know that you've met the learning intentions of our lesson. Look back in your mathematician's journal for evidence that fits each of the four success criteria. You might star, highlight, or label it.

Students reread their work and make notes with their colored pencils.

> Finally, take a sticky note and put your initials on it. As you put away your mathematician's journal, place your sticky note on the target. Remember, the bull's-eye means "I've got it!" The middle ring is "I need a little more time. I'm just starting to get it." The outer ring is "I'm stuck. I don't understand." Tomorrow, I will meet with you to dig deeper into where you are with these learning intentions.

Figure 3.5 shows how Ms. Mills made her planning visible so that she could then provide an engaging and rigorous learning experience for her learners.

Teaching Takeaway

Providing opportunities for learners to think aloud provides opportunities for students to model their thinking and receive feedback.

EFFECT SIZE FOR SELF-VERBALIZATION AND SELF-QUESTIONING = 0.55

EFFECT SIZE FOR SETTING STANDARDS FOR SELF-JUDGEMENT = 0.62

Video 9
Learning Intentions and Success Criteria Throughout a Lesson

https://resources.corwin.com/vlmathematics-3-5

Ms. Mills's Teaching for Clarity PLANNING GUIDE

ESTABLISHING PURPOSE

1

What are the key content standards I will focus on in this lesson?

Virginia Mathematics Standards of Learning

4.2c. The student will identify the division statement that represents a fraction, with models and in context (VDOE, 2016).

Mathematical Process Standards:

- Mathematical communication
- Mathematical problem solving
- Mathematical representations

2

What are the learning intentions (the goal and *why* of learning stated in student-friendly language) I will focus on in this lesson?

- Content: I am learning to understand the connections among representations of fractions.
- Language: I am learning to understand the language that makes meaning of fraction notation.
- Social: I am learning to understand how to listen and respond to my peers' ideas in ways that move us all forward as learners.

3

When will I introduce and reinforce the learning intention(s) so that students understand it, see the relevance, connect it to previous learning, and can clearly communicate it themselves?

- Post the learning intentions.
- Engage in turn and talk after activating prior knowledge.

- Make connections during think-alouds, visualizing, conferences, and sharing.
- Use mathematician's journal notes and final reflection.

SUCCESS CRITERIA

4 What evidence shows that students have mastered the learning intention(s)? What criteria will I use?

I can statements:

- I can represent fractions using multiple representations.
- I can explain the connections among these representations.
- I can use these representations to prove fractions are equivalent.
- I can communicate my process, my representations, and my mistakes using fraction language: number of pieces (numerator), size of piece (denominator), and same size share (equivalent fractions).

5 How will I check students' understanding (assess learning) during instruction and make accommodations?

Formative Assessment Strategies:

- Open conference chart notes
- Mathematician's journal
- Bull's-eye self-evaluation

Differentiation Strategies:

- Differentiate the content by interest: anchor problem and extension problems.
- Differentiate the process by interest: choice of materials and partners or alone.

INSTRUCTION

6   What activities and tasks will move students forward in their learning?

- Fraction Graffiti activity
- Equalizing conservation problems with CRA representations
- Mathematician's journal reflection
- Bull's-eye self-evaluation

7   What resources (materials and sentence frames) are needed?

STAR anchor chart

Language frames

Mathematician's journals

Cubes

Cuisenaire rods

Fraction pieces

Graph paper

Open number lines and whiteboard markers

Colored pencils

Calculators

8   How will I organize and facilitate the learning? What questions will I ask? How will I initiate closure?

Instructional Strategies:

- STAR strategy with think-alouds and visualizing
- Anticipate, monitor, select, sequence, and connect students' CRA strategies
- Turn and talk

Scaffolding Questions:

- How could these cubes represent what we know?

- How will you know when each panda has an equal share?

Extending Questions:

- What if there were five pandas in the first zoo? How would this change the solution?

- What if there were nine pandas in the second zoo?

Connecting Questions:

- How are the representations similar? Where do you see $\frac{1}{4}$ and $\frac{2}{8}$ in each?

- How is doubling related to equivalence in this problem?

- What representations were efficient, flexible, or accurate? What representations were not? Why?

- How is 1 ton divided by 4 pandas related to $\frac{1}{4}$? And 2 tons divided by 8 pandas related to $\frac{2}{8}$?

Self-Reflection and Self-Evaluation for Closure:

- Mathematician's journal reflection

- Bull's-eye self-evaluation

 This lesson plan is available for download at **resources.corwin.com/vlmathematics-3-5**.

Figure 3.5   Ms. Mills's Conceptual Understanding Lesson on Division as Fractions

Ms. Campbell and the Volume of a Rectangular Prism

This unit on volume continues the development of geometric measurement thinking Ms. Campbell's students began in earlier grades. Ms. Campbell knows that much of her students' current experience with volume is in the context of "filling," in which students measure the volume of water or soil for experiments in their science class using materials that take the shape of the container. They have used graduated cylinders to measure and compare the amount of water given to plants and use milliliters as a unit of measure for this work. To develop a conceptual understanding of the volume formula, students need to understand volume as "packing," where unit cubes are used to fill the space with no gaps and no overlaps. This highlights the "cubic units" measure used for volume and the connection that *unit* will have to the procedure used to calculate the volume of right rectangular prisms.

What Ms. Campbell Wants Her Students to Learn

Ms. Campbell relies on her state standards to guide the mathematics learning of her fifth grade students. Her years of experience have taught her the variability of students' prior knowledge and the previous experiences of fifth grade learners, and she has a good grasp of the key elements of the content. This unit on volume is placed early in the year and builds students' algebraic reasoning as they develop a deep understanding of the volume formula in the context of packing a shape with unit cubes rather than filling an object with liquid. This first lesson is designed to bridge from students' filling concept of volume from science class to the packing concept important in the mathematics standards.

MATHEMATICS CONTENT AND PRACTICE STANDARDS

5.MD.C. Understand concepts of volume and relate volume to multiplication and to addition.

3. Recognize volume as an attribute of solid figures and understand concepts of volume measurement.

   a. A cube with side length 1 unit, called a "unit cube," is said to have "one cubic unit" of volume and can be used to measure volume.

   b. A solid figure which can be packed without gaps or overlaps using *n* unit cubes is said to have a volume of *n* cubic units.

(Continued)

EFFECT SIZE FOR STRATEGY TO INTEGRATE WITH PRIOR KNOWLEDGE = 0.93

EFFECT SIZE FOR IMAGERY = 0.45

Teaching Takeaway

Learners come to our classrooms with different experiences, interactions, and vocabulary. Taking this into consideration with new learning is an essential part of teacher clarity.

EFFECT SIZE FOR TEACHER CLARITY = 0.75

EFFECT SIZE FOR INTEGRATED CURRICULA PROGRAMS = 0.47

(Continued)

4. Measure volumes by counting unit cubes, using cubic cm, cubic in, cubic ft, and improvised units.

Ms. Campbell is helping her learners develop the following Standards for Mathematical Practice:

- Use appropriate tools strategically.
- Attend to precision.

Ms. Campbell's students will work on these practices by using various tools for measuring volume and focusing on the precision required to pack figures into a rectangular prism with no gaps or overlaps.

Learning Intentions and Success Criteria

Ms. Campbell has been working with the other math teacher on her team to develop clearer learning intentions and success criteria for their lessons. These teachers have realized that students are more engaged with their learning when they understand the purpose of the math lesson in their own terms. Rather than just writing the standards on the board, Ms. Campbell now explains her learning intentions in terms of the content, language, and social focus of each lesson. For today's lesson, she wants to focus on volume as the space inside a solid figure and begin to use the vocabulary associated with measuring volume by packing.

EFFECT SIZE FOR
CONCENTRATION/
PERSISTENCE/
ENGAGEMENT
= 0.56

Teaching Takeaway

Working together with colleagues to ensure that we are having the greatest impact on student learning is one of the key mindframes for teaching mathematics in the Visible Learning classroom.

Content Learning Intention: I am learning that volume is the amount of space inside a solid figure.

Language Learning Intention: I am learning to use the mathematics language to describe volume (i.e., *capacity, cubic units, packing, gaps, overlaps*).

Social Learning Intention: I am learning how to record and explain my work clearly for my classmates.

The success criteria of the lesson serve as markers for the students to measure their own progress toward meeting the goal of the lesson.

☐ I can explain what volume is.

☐ I can *still* identify and describe a right rectangular prism.

☐ I can count or measure the amount of space inside a right rectangular prism.

☐ I can use volume to describe and compare the capacity of a solid shape.

Activating Prior Knowledge

Today's lesson is designed to help students understand what volume is from a mathematical perspective and to begin to transfer their ideas of volume as filling to the idea of volume as packing. Each pair of students begins class with a box that is a small right rectangular prism. Ms. Campbell begins the lesson.

Teaching Takeaway

Using the term *still* in our success criteria allows students to see learning as a process and the assimilation of new learning with prior learning.

Sources: tropper2000/iStock.com, simplyzel/iStock.com, and sorendls/iStock.com (left to right)

Alright folks, let's get our minds on math! Today you will be using your mathematical minds to think about the volume of shapes called right rectangular prisms. Each of the boxes on your table is a right rectangular prism. Take 2 minutes with your partner to look at your box and record a mathematical statement you can make about the box in your journals. Use your math vocabulary in your statements; the word wall might help you. For example, what can you say about the vertices of this object?

MS. CAMPBELL'S OBSERVATION/CONFERENCE CHART

Pair	Parts of Box	Why Right Rectangular Prism?	Relative Size	Volunteer at Right Time	Notes for Future Teaching
A					
B					
C					
D					

Figure 3.6

EFFECT SIZE FOR VOCABULARY INSTRUCTION = 0.62

Ms. Campbell listens while student pairs talk about their boxes. She is pleased to see students using the word wall to help them; this is her first year making a word wall, an idea she found when she visited a third grade colleague's classroom last year. As Ms. Campbell circulates the room, she listens for groups who are describing the parts of the box (faces, edges, vertices), for groups who are explaining why the box is a right rectangular prism (parallel faces, rectangles connecting the faces, right angles), and for groups who are talking about the relative size of the box. She anticipated these three groups of responses and plans to sequence sharing in this order. Her record-keeping sheet (Figure 3.6) serves as a reminder of her observations as well as documentation of student thinking in this part of the lesson.

EFFECT SIZE FOR RECORD KEEPING = 0.52

After 2 minutes have passed, Ms. Campbell asks the students to give her their attention for a class discussion.

EFFECT SIZE FOR STRATEGY MONITORING = 0.58

As I observed you work, I saw three different kinds of statements in your journals. Some teams were naming the parts of the box. Other teams were explaining why the box is a right rectangular prism. Other teams were talking about the size of the box. Take 30 seconds with your partner to think about which category your observation falls in.

I'd like to hear from the teams who named the parts of the
box first. If your statements are about the vocabulary of
these shapes, please raise your hand to show me you're ready
to share.

Ms. Campbell likes this strategy because it gives her a quick assessment
of the match between her notes about student statements and those
who volunteer. She can monitor student self-assessment by noting the
match on her observation record.

"David, please read your statement about the box," she says.

"Our box has six faces and eight corners," David replies.

Ms. Campbell says, "Thank you, David. I heard David say that his box
has six faces and eight corners. Look at your own box. Does your box
have six faces?" She watches the class count, looking carefully for students who are counting the wrong part of the shape so she can give a
brief vocabulary review at a later time.

> EFFECT SIZE FOR
> FEEDBACK = **0.70**

> EFFECT SIZE
> FOR PROVIDING
> FORMATIVE
> EVALUATION = **0.48**

"Give me a thumbs up if your box also has six faces," she continues.
Ms. Campbell sees that everyone has given a thumbs up and continues:
"Is there another mathematical term we can use for corners? Does another
group have a description which uses a different vocabulary term?"

Conceda shares, "We said our box has six faces or sides and eight ver-
texes." Ms. Campbell smiles and thanks her. "Great. You're right that
vertex is our mathematical term for corner and we say *vertices* when the
word is plural. Class, take a moment to check if your figure also has
eight vertices. Give me thumbs up if you agree and thumbs down if
you don't."

Seeing thumbs up around the room, Ms. Campbell continues by asking
a group to share why they think the shape is a right rectangular prism.
Jose shares for his team: "We saw that the top and the bottom of our box
were the same shape, rectangles, and all of the angles are right angles."
"Is that true for all of the boxes?" Ms. Campbell asks. While students
examine their boxes, she continues. "This is what makes the shape a
right rectangular prism. The bases, the top and bottom of the shape, are
the same rectangle and all the corners are 90 degrees. Does everyone
agree that all the boxes are right rectangular prisms?" She provides her

learners an additional opportunity to review the essential characteristics of a right rectangular prism and says, "Take a quick second and take turns identifying what makes a right rectangular prism a right rectangular prism."

The teacher moves the lesson along. She will return to these ideas when students study classification of geometric figures later in the year. For now, she has established that the shapes are all the same and that students will understand the term *right rectangular prism* sufficiently when it appears in the day's learning intentions and success criteria. She continues the lesson by introducing the learning intentions and success criteria for the day.

Ms. Campbell says, "For today's lesson, you will be rotating to at least four of the six stations in the room and you will measure the volume of a box. *Volume* is our mathematical term for how much capacity a solid figure has—how much material it will take to fill up the box. I am going to model what you will do at each station now." Ms. Campbell picks up a tray, which she places on the table along with her box. She picks up the containers on the tray one at a time to show the class the materials on the tray.

> I have cubes in two different sizes; some are bigger and some are smaller. I have some two-color counters. I have rice with a scoop, and I have some marbles. At each station, your job is to choose one of the materials and measure how much of that material will fill the box at the table. Let me show you using this box.

Ms. Campbell picks up her box and some of the smaller cubes. She begins to pack the centimeter cubes neatly into the box.

> I'm filling the box with small cubes. See how I'm packing them in here? I'm keeping them lined up so there are no empty spaces, no gaps, in my filling. I'm going to come as close to the sides of the box as I can. Once I get one layer filled in, I'll build another layer until the box is as full as I can make it. Then I'll count the cubes.

She continues to model the use of each resource, pouring in scoops of rice while counting, filling with marbles then counting the marbles, and filling with two-color counters.

> In a moment, one person from your table will come get your tray of materials to measure volume. You will start your work with the box at your station. Choose one material and use it to measure the volume of your box. Record your work on the recording sheet at your station. You saw me fill my box with small cubes. Here is how I would fill in the recording sheet.

Ms. Campbell models filling in the cells for the measurement she made.

> Notice that I included units on the amount to fill the box. Think about the best unit to use for each measurement you take.

Box A

Group	Measuring Material	Amount to Fill the Box
Teacher	Small cubes	16 small cubes

When you finish with your measurement and recording, your team should move to a second station and pick a different material to use to measure the volume of the box at that station. Remember our class guidelines for working in rotating stations. Today, there can be no more than two groups at any one station, and I expect you to visit at least four stations during our work time. Judy, would you tell us what you understand your team's work is when you are at each station?

EFFECT SIZE FOR MANIPULATIVE MATERIALS ON MATH = 0.30

Teaching Takeaway

Although manipulatives have an average effect size less than the hinge point of 0.40, manipulatives help learners make meaning of abstract concepts through concrete representations.

EFFECT SIZE FOR DIRECT/ DELIBERATE INSTRUCTION = 0.60

EFFECT SIZE FOR STRONG CLASSROOM COHESION = 0.44

Strong classroom cohesion exists when there is a sense that all (teachers and students) are working toward positive learning gains.

Judy says, "We start at our own table. We have to pick a material and then measure the amount that fills the box and record it."

The teacher replies, "Thank you, Judy. Would anyone like to add details to what Judy said? Cecelia?"

Cecelia says, "When we measure with cubes, we have to fill them in all the way. We want the box to be as full as we can make it. We should use units when we record our measurements."

"Thank you, Cecelia," Ms. Campbell replies. "Now, who can tell us the guidelines for rotating to stations? Alex, thank you for volunteering."

Alex says, "We have to go to four stations and there can only be two groups at one station at a time. Do we have to measure with something different each time?"

"Thank you, Alex," Ms. Campbell says. "Yes, you'll be measuring with a different filling each time. It sounds like we are ready to begin. The trays of materials are on the counter. Please send one person to get your tray and you can get started. You have about 20 minutes to make your measurements."

EFFECT SIZE FOR
SUMMARIZATION
= 0.79

Scaffolding, Extending, and Assessing Student Thinking

As the class begins to work, Ms. Campbell watches to be sure each group is on track with their first measurement. As she circulates the room, she checks that groups are using a variety of measuring materials and recording their measurements with units on the recording sheets. When she reaches Alex's table, she says "I see you are using marbles to measure here. How do you know you have as many marbles as possible in the box?" "We poured them in and I used my hand to push off the extra," Alex says, as he motions smoothing the top of the box with his hand to eliminate marbles sitting above the top of the box. Ms. Campbell notices that the marbles are not fully packed into the box and asks the group to make sure they have as many marbles as possible in the box. Nancy picks up the box and the marbles settle as she moves it. Nancy reaches for more marbles as the teacher moves to the next table.

Ms. Campbell asks another group what they notice about their measurements. Kiernan replies, "The marbles are in between the size of the two cubes, but it doesn't take as many marbles to fill up the box." Kiernan lines up a small cube, a marble, and a large cube in size order to show what he means. Ms. Campbell suggests, "Think about how marbles pack the box and how cubes pack the box. Is there a difference?" She wants the group to think about the importance of packing with no gaps and no overlaps; this is why the marbles are included in this activity.

After students complete their measurements, Ms. Campbell brings their attention back to the front of the room. She asks one student from each table to bring their box and its measurement recording sheet to the front of the room.

Ms. Campbell selects two boxes, one clearly larger than the other. She asks the class, "Which box is larger?" They laugh at this baby-ish question and point to the larger box. "How do you know it's larger? How much larger?" the teacher asks. When Ms. Campbell asks these questions, the class quiets.

> EFFECT SIZE FOR
> QUESTIONING
> = **0.48**

Ms. Campbell continues, "Would it help to know some of the measurements for these boxes? Talk with your table group about what information you would like." After waiting a moment for conversation, Ms. Campbell calls on Kim and says, "What information would your group like, Kim?" Kim asks for the number of small cubes that fill each box. Ms. Campbell tells her that 48 small cubes fill Box B and 82 small cubes fill Box D.

"Then Box D is larger because it takes more small cubes to fill it up," says Kim.

Ms. Campbell asks, "Cecelia, do you agree? Why or why not?"

"I agree. I could subtract to find out how much bigger 82 is than 48," replies Cecelia.

The teacher says, "Thank you, Cecelia. I like that you shared a strategy for finding how much bigger Box D is than Box B. Class, we know it took 48 small cubes to fill Box B. It took 6 large cubes to fill the same box. What does that tell us? Turn to your shoulder partner and share your thinking."

> EFFECT SIZE
> FOR CLASSROOM
> DISCUSSION = **0.82**

After allowing a minute for discussion, Ms. Campbell calls on Nancy to share.

> We think it's like measuring area. When you measure in square inches, the units are little and it takes a lot of them to cover something up. When you measure in square feet, the units are bigger so it doesn't take as many. Maybe it takes only six large cubes because they're bigger so it doesn't take as many to fill up the box.

The teacher says, "Thank you, Nancy. Who would like to respond to Nancy's statement? Emily?" Emily shares the following:

> My group used small cubes to try to make a large cube. We used eight small cubes and it was almost as big as the large cube. That means we could take out eight small cubes and put in one large cube. Since 48 divided by 8 is 6, we agree with Nancy.

Video 10
Questioning and Discourse to Clarify and Deepen Understanding

https://resources.corwin.com/ vlmathematics-3-5

Ms. Campbell continues the discussion by comparing other box pair and measurement values. This time, she focuses on measuring with rice, connecting to the idea of filling from science class. Her students observe that it is more difficult to compare volume when measuring the rice in scoops because they aren't sure the scoops are the same size. When they compare the size of two boxes with marbles, they realize that there are gaps because the marbles don't pack together tightly and the students wonder if that impacts the measurement of volume.

"I think it does," Nancy says. "We want to know how much space is in the box and we're only counting the part of the space with marbles, not the empty spaces around them."

Alex raises his hand and continues, "I disagree. As long as the marbles fit together, we can compare. The empty space doesn't matter."

"But we noticed that our marbles didn't always fit together the same way," Nancy shares. "If we shook the box, we could put more in. When will it stop?"

Teaching for Clarity at the Close

Ms. Campbell brings the class back to the success criteria for the lesson as students prepare to self-assess their progress:

- I can explain what volume is.
- I can *still* identify and describe a right rectangular prism.
- I can count or measure the amount of space inside a right rectangular prism.
- I can use volume to describe and compare the capacity of a solid shape.

EFFECT SIZE FOR
EVALUATION AND
REFLECTION = 0.75

Students paste the list into their journal and use colored pencils to mark their progress. Green in the circle means "I'm ready to teach this to someone else." Yellow in the circle means "I think I understand this but have some questions." Red in the circle means "I don't understand this yet." However, Ms. Campbell wants learners to support their evaluation of learning.

Video 11
Practicing Evaluating and
Giving Feedback

*https://resources.corwin.com/
vlmathematics-3-5*

> Next to your progress circles, I would like you to explain your reason for marking the success criteria green, yellow, or red. For example, what are you ready to teach and how would you do it? For red, you might list questions you still have or identify specific things you don't yet understand.

> Tomorrow, our work together will focus on measuring volume using cubic units, like cubic inches or cubic centimeters. Does that help you think about the challenge of measuring with the different units we used today?

Figure 3.7 shows how Ms. Campbell made her planning visible so that she could then provide an engaging and rigorous learning experience for her learners.

Ms. Campbell's Teaching for Clarity PLANNING GUIDE

ESTABLISHING PURPOSE

1

What are the key content standards I will focus on in this lesson?

Content Standards:

> *5.MD.C. Understand concepts of volume and relate volume to multiplication and to addition.*

3. *Recognize volume as an attribute of solid figures and understand concepts of volume measurement.*

    a. *A cube with side length 1 unit, called a "unit cube," is said to have "one cubic unit" of volume and can be used to measure volume.*

    b. *A solid figure which can be packed without gaps or overlaps using n unit cubes is said to have a volume of n cubic units.*

4. *Measure volumes by counting unit cubes, using cubic cm, cubic in, cubic ft, and improvised units.*

Standards for Mathematical Practice:

- *Use appropriate tools strategically.*

- *Attend to precision.*

2

What are the learning intentions (the goal and *why* of learning stated in student-friendly language) I will focus on in this lesson?

- *Content: I am learning that volume is the amount of space inside a solid figure.*

- *Language: I am learning to use the mathematics language to describe volume (i.e., capacity, cubic units, packing, gaps, overlaps).*

- *Social: I am learning how to record and explain my work clearly for my classmates.*

3 When will I introduce and reinforce the learning intention(s) so that students understand it, see the relevance, connect it to previous learning, and can clearly communicate it themselves?

- *Look at boxes and define a right rectangular prism, then share learning intentions before beginning the activity.*

SUCCESS CRITERIA

4 What evidence shows that students have mastered the learning intention(s)? What criteria will I use?

I can statements:

- *I can explain what volume is.*

- *I can still identify and describe a right rectangular prism.*

- *I can count or measure the amount of space inside a right rectangular prism.*

- *I can use volume to describe and compare the capacity of a solid shape.*

5 How will I check students' understanding (assess learning) during instruction and make accommodations?

Formative Assessment Strategies:

- *Observe student collaboration and discussion.*

- *Review student self-assessments.*

Differentiation Strategies:

- *Differentiate the process by interest: choice of shapes to fill and fill materials.*

INSTRUCTION

6 **What activities and tasks will move students forward in their learning?**

- *Observing the boxes to build vocabulary*
- *Measuring volume station rotations*
- *Closing self-assessment and task*

7 **What resources (materials and sentence frames) are needed?**

Boxes to measure

Measuring materials: centimeter cubes, inch cubes, marbles, rice, two-color counters (sufficient amounts for the boxes)

Recording sheet for box measurements

Success criteria lists for journals

8 **How will I organize and facilitate the learning? What questions will I ask? How will I initiate closure?**

Instructional Strategies:

- *Anticipate, monitor, select, sequence, and connect student responses observing the boxes*
- *Facilitate group measurement task rotation*

Scaffolding Questions:

- *What measurement tool are you using at this station?*
- *How will you know you have filled the box as much as you can?*

Extending Questions:

- *What pattern do you notice?*
- *How are the quantities changing?*

Connecting Questions:

- *How does this measuring tool compare to another one you have used?*

- *How do your measurements for this box compare to the measurements others have made?*

Self-Reflection and Self-Evaluation for Closure:

- *Color-coded assessment of success criteria*

 This lesson plan is available for download at **resources.corwin.com/vlmathematics-3-5**.

Figure 3.7　Ms. Campbell's Conceptual Understanding Lesson on the Volume of a Rectangular Prism

Reflection

These three examples from Ms. Buchholz, Ms. Mills, and Ms. Campbell exemplify teaching mathematics for conceptual understanding. As in the previous chapter, these three teachers selected a different approach or combination of approaches from the other two classrooms.

Using what you have read in this chapter, reflect on the following questions:

1. In your own words, describe what teaching for conceptual understanding looks like in your mathematics classroom.

2. How does the Teaching for Clarity Planning Guide support your intentions in teaching for conceptual understanding?

3. Compare and contrast the approaches to teaching taken by the classroom teachers featured in this chapter.

4. Consider the following statement: *Conceptual understanding occurs at the surface, deep, and transfer phases of learning.* Do you agree or disagree with the statement? Why or why not? How is this statement reflected in this chapter?

5. How did the classroom teachers featured in this chapter adjust the difficulty and/or complexity of the mathematics tasks to meet the needs of all learners?

TEACHING FOR PROCEDURAL KNOWLEDGE AND FLUENCY

4

CHAPTER 4 SUCCESS CRITERIA:

(1) I can describe what teaching for procedural knowledge in the mathematics classroom looks like.

(2) I can apply the Teaching for Clarity Planning Guide to teaching procedural knowledge.

(3) I can compare and contrast different approaches to teaching for procedural knowledge with those of teaching for conceptual understanding and application.

(4) I can give examples of how to differentiate mathematics tasks designed for procedural knowledge.

Procedural knowledge is the ability to select, use, and transfer mathematics procedures in problem solving. With procedural knowledge, learners know when one procedure is more appropriate than another one for a particular problem.

In mathematics, you have to be able to solve problems and reason quantitatively. The successful teaching and learning of mathematics may involve the execution of procedures and quantitative reasoning that yield an expression, value, or set of values. Acquiring and consolidating **procedural knowledge**—which is the ability to select, use, and transfer mathematics procedures in problem solving—is a necessary aspect of mathematics if learners are to have the appropriate tools for taking on the next challenge in their learning progression. As we make our final visit to our three featured teachers, we want to take a look at how each teacher created learning experiences that allowed students to learn the necessary procedural skills and progress toward fluency with those skills. Also, you'll notice the adjustments each teacher made to the learning experience so that learners at the surface, deep, and transfer phases of learning could all engage in the mathematics task. And as before, you'll see how Ms. Buchholz, Ms. Mills, and Ms. Campbell were able to differentiate the rigor of the mathematics tasks.

Ms. Buchholz and Fluent Division Strategies

Ms. Buchholz's students are deeply engaged in answering the essential questions of their multiplication and division unit. Over the last few weeks, they solved five types of multiplication problems: equal grouping, array/area, rate/price, multiplicative comparison, and combinations (National Research Council, 2009). This work developed their conceptual understanding of the meaning of multiplication as well as many problem-solving strategies for multiplication.

EFFECT SIZE FOR DELIBERATE PRACTICE = **0.79**

They gathered multiplication strategies on an anchor chart and evaluated each for efficiency, accuracy, and flexibility. Their procedural knowledge of multiplication strategies is deep. Students' levels of procedural fluency with multiplication basic facts vary from surface to deep to transfer learning. As they engage in **spaced practice** to further develop their fluency, Ms. Buchholz moves their learning forward by examining the meaning of division and the relationship between multiplication and division.

Spaced practice is practice that occurs over time rather than in a single setting or practice session.

During the last 3 days, students used their growing mathematical toolbox to solve the same five types of problems but presented as division. From that work, the students started to answer the essential question,

EFFECT SIZE FOR SPACED VS. MASS PRACTICE = **0.60**

What is division? by saying, "Division is partitioning, sharing, and the inverse of multiplication." Students are also at different levels of understanding the concept of division. Ms. Buchholz knows that her students will not all reach mastery at the same time nor through the same experiences. She carefully plans today's work to allow time to work on multiplication fluency, conceptual understanding of division, and procedural knowledge of division strategies. Her students have a list of tasks to complete. The must-do task aligns with the lesson's learning intentions and provides targeted opportunities for students to demonstrate the success criteria. The may-do tasks provide optional, additional opportunities for students to deliberately practice areas of need related to the unit's essential questions (Thunder & Demchak, 2017).

The must-do task asks students to use the definition of procedural fluency (efficiency, accuracy, and flexibility) (NCTM, 2014) to evaluate four division strategies and to test representations for each strategy.

> EFFECT SIZE FOR
> EXPLICIT TEACHING
> STRATEGIES = 0.57

Evaluating Fluency

How can I demonstrate procedural fluency for the problem 63 ÷ 9?

Which division strategies should I use? Which representations should I use?

Mathematicians value being fluent in problem solving. They evaluate strategies using three criteria:

- Accuracy: Will you always get the right answer using this strategy?
- Efficiency: Is this a relatively quick, easy-to-keep-track-of strategy?
- Flexibility: Does this strategy work with many different numbers?

We need to know which of our four division strategies we should spend time deliberately practicing: sharing or dealing out, repeated subtraction, multiplying up, and partial quotients.

We also need to know which representations work well with each strategy: manipulatives (cubes), pictures, number lines, and equations. A good representation supports the accuracy, efficiency, and flexibility of the strategy and helps other mathematicians understand why the strategy works.

Decide which division strategies will support fluency. Decide which representations work well for each strategy. You may work alone or with classmates.

Teaching Takeaway

Anchor charts provide opportunities for learners to self-regulate their learning. They can simply monitor their learning by finding anchor charts around the room to support their thinking.

The students had previously developed and shared four division strategies as they solved the contextualized division problems. These strategies are now displayed on an anchor chart for division with examples and strategy names (Parrish, 2014). Naming the strategies with common language enables the students to talk and make decisions about the strategies.

Some may-do tasks allow time for students to deliberately practice their multiplication fluency by creating skip-counting number charts of multiples and by playing two games (the Product Game and Multiplication Bingo). One may-do task engages students in extending their conceptual understanding of multiplication: the Mysterious Multiplying Jar Problem (Anno, 1983). Other may-do tasks engage students in bolstering their conceptual understanding of division by solving additional contextualized division problems. Through this mixture of must-do and may-do tasks, Ms. Buchholz will meet her students where they are in their growth toward mastery of the learning intentions.

What Ms. Buchholz Wants Her Students to Learn

Although her students will work on may-do tasks that address additional standards from throughout the unit, the must-do task targets standards that resonate with the essential questions for today's lesson: *What is division? How are division and multiplication related?*

Teaching Takeaway

Meet learners where they are in their learning progression. Let their learning define them and not be a label.

INDIANA ACADEMIC STANDARDS

3.C.5. Multiply and divide within 100 using strategies, such as the relationship between multiplication and division (e.g., knowing that 8 × 5 = 40, one knows 40 ÷ 5 = 8), or properties of operations.

3.AT.2. Solve real-world problems involving whole number division within 100 in situations involving equal groups, arrays, and measurement quantities (e.g., by using drawings and equations with a symbol for the unknown number to represent the problem).

3.AT.5. Determine the unknown whole number in a division equation relating three whole numbers.

> **Ms. Buchholz is helping her learners develop the following Standards for Mathematical Practice:**
>
> - Reason abstractly and quantitatively.
> - Look for and express regularity in repeated reasoning.

Teaching Takeaway

Using guiding or essential questions enhances the clarity around learning intentions and success criteria.

Learning Intentions and Success Criteria

Several weeks into the school year, Ms. Buchholz's students now expect to know, discuss, and reflect on the learning intentions and success criteria for each lesson. They know Ms. Buchholz creates three types of learning intentions. The first communicates to her students the new mathematical content they are learning. The second delineates the new language they will use. This deliberate focus on language learning supports the many English language learners in Ms. Buchholz's class as well as all of the students who are making sense of new academic vocabulary and sentence structures. The third learning intention makes explicit the social learning process that is ongoing as a classroom community.

> *Content Learning Intention*: I am learning to evaluate fluent and nonfluent division strategies based on the representation and the numbers.
>
> *Language Learning Intention*: I am learning how the language of division (*partitioning*, *sharing*, and the *inverse of multiplication*) can be used to explain why a strategy will always work.
>
> *Social Learning Intention*: I am learning to reflect on the decisions we make as learners and how they help us grow as individuals and a community of learners.

Ms. Buchholz then uses the learning intentions to create the success criteria that she and the students use to know whether they mastered the learning intentions.

☐ I can use and evaluate efficient division strategies.

☐ I can use and evaluate efficient mathematical models to represent division strategies.

☐ I can explain why a division strategy will always work using division language (*partitioning*, *sharing*, and the *inverse of multiplication*).

☐ I can explain the inverse relationship between multiplication and division.

The teacher unpacks the content learning intention to create the first two success criteria. The language learning intention is translated into the last two success criteria. The emphasis on *evaluate* and *explain* in the success criteria reflects the social learning intention. Ms. Buchholz plans to introduce the learning intentions and success criteria after she activates their prior knowledge and before she explains their must-do and may-do tasks.

EFFECT SIZE FOR SELF-VERBALIZATION AND SELF-QUESTIONING = 0.55

Activating Prior Knowledge

EFFECT SIZE FOR PRIOR ACHIEVEMENT = 0.55

Ms. Buchholz's students are gathered around her whiteboard and counting together by fours: "4, 8, 12, 16, 20, 24, 28, 32, 36, 40." The teacher pauses and asks, "How many times have we counted by four?" Many thumbs pop up. Ms. Buchholz records her question with mathematical notation in two ways: $4 \times \underline{} = 40$ and $40 \div 4 = \underline{}$.

"Ten," Leo explains, "We have 10 numbers written down."

"Is there another way to confirm Leo's conjecture?" Ms. Buchholz asks.

EFFECT SIZE FOR QUESTIONING = 0.48

"Ten groups of four is 40 so if we've counted to 40, we've counted 10 fours," Hadley adds.

Ms. Buchholz completes the equations: $4 \times 10 = 40$ and $40 \div 4 = 10$.

"Let's continue." Ms. Buchholz moves to the next row and records the students' count: "44, 48, 52, 56, 60, 64, 68, 72, 76, 80."

"Now how many fours have we counted?" Ms. Buchholz asks. "Turn and tell a person sitting next to you." She records $4 \times \underline{} = 80$ and $80 \div 4 = \underline{}$. She hears students explaining that the two rows of

10 mean they have counted 20 fours. "How many fours have we counted?" Ms. Buchholz again asks the whole group. "Twenty," they respond. Again, she completes the equations: $4 \times 20 = 80$ and $80 \div 4 = 20$.

Next, Ms. Buchholz draws a box under 64 and says, "Based on the pattern, what number will go here? Share your conjecture with a partner and defend it."

> 4, 8, 12, 16, 20, 24, 28, 32, 36, 40
>
> 44, 48, 52, 56, 60, 64, 68, 72, 76, 80
>
> ☐

Again, Ms. Buchholz listens in as pairs discuss their ideas. Most students defend 104 but one pair thinks it is 74. This mistake makes sense to Ms. Buchholz; on a number chart, 74 is always directly below 64.

"What numbers did you decide could go here?" Ms. Buchholz asks. She purposefully allows for multiple answers in her question. Bringing mistakes and misconceptions to light through discussion is important to all students' learning. Often other students are thinking the same way or may in the future. In Ms. Buchholz's classroom, her students know reasoning through why a mistake is wrong but makes sense is as valuable as defending why a correct answer is right. This is all part of Ms. Buchholz's core belief that everyone is a student and everyone is a teacher.

EFFECT SIZE FOR ASSESSMENT-CAPABLE VISIBLE LEARNERS = **1.33**

Two answers are offered: 104 and 74. Ms. Buchholz records both and asks, "Who will defend 104?"

Frances begins, "We predict 104 goes in the box. We checked it two ways. We counted on by fours from 80 six more times to get to the box: 84, 88, 92, 96, 100, 104. We also conjecture there's a skip-counting pattern in the columns that counts by 40s, like 40 and 40 is 80. So we counted by forties starting at 24, which is above 64: 24 and 40 is 64, 64 and 40 is 104."

"Who will defend 74?" Ms. Buchholz asks.

"We thought 74 because that's how hundred charts work—under 64 is 74. But now I think it's 104. I see the counting by 40 pattern Frances described," volunteers Monique.

Teaching Takeaway

Through teacher-led dialogue, we can both model our thinking while at the same time make student thinking visible.

"And 74 wouldn't make sense because the end of the second row is 80. The numbers in the next row should all be greater than 80. I didn't think about that before," adds Sabina.

"How many fours does it take to get to 104?" Ms. Buchholz asks. She records two equations again: $4 \times \_\_\_ = 104$ and $104 \div 4 = \_\_\_\_$. She waits until she sees most students put their thumbs up. First, Ms. Buchholz collects all answers: 26, 27, 46. Then she asks students to defend each answer. As is typical, students who shared incorrect answers self-correct and explain why their initial responses made sense but were wrong. The discussion ends with Ms. Buchholz recording $4 \times 26 = 104$ and $104 \div 4 = 26$.

Through this classroom routine, a counting math talk, Ms. Buchholz is able to activate students' prior knowledge. Her students are thinking multiplicatively. They are thinking flexibly about the relationship between multiplication and division. They are making connections among multiplicative notation and language.

Now, Ms. Buchholz introduces the learning intentions and success criteria for the day. She says, "Earlier in our work, we gathered multiplication strategies and then evaluated them. That's how we made this anchor chart and added to your mathematical toolbox. Today, we're going to do similar work but with the division strategies we've gathered from your problem solving."

Students talk in pairs about the learning intentions and success criteria. Many students are talking to their partners in their heritage language. This is an important first step in making sense of academic language, and Ms. Buchholz's students know they can always talk in their heritage language to help them make sense of new vocabulary, academic sentence structures, and problem contexts.

Several students share they are not sure what the phrase *inverse of multiplication* means. Others respond by pointing out Ms. Buchholz's recording with two equations during the counting math talk. Geneva says, "You can think of each problem as division or multiplication." Ms. Buchholz asks them to be on the lookout for more examples and ways to explain this big idea as they work.

Next, Ms. Buchholz introduces the must-do task. She takes down the word *fluency* from the word wall and displays the back of the card with three images (Thunder & Demchak, 2016).

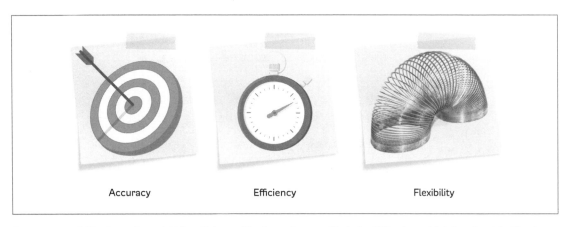

| Accuracy | Efficiency | Flexibility |

Sources: art-sonik/iStock.com (target), Mykyta Dolmatov/iStock.com (stopwatch), shellystill/iStock.com (slinky), and rami_ba/iStock.com (sticky notes)

"Talk with a partner to remember what it means when a strategy helps us to be fluent problem solvers. Use these pictures and words to help you," she says.

EFFECT SIZE
FOR STRATEGY
MONITORING = 0.58

Next, Ms. Buchholz passes out the task and reads it aloud. She says, "Talk with a partner. How are you making sense of your must-do task today? What will you do? What will your product be?" After students talk, they ask clarifying questions and their peers respond. Ms. Buchholz then explains today's lesson.

> For today's must-do task, I assigned partners. You will work together to engage in rough-draft talk and rough-draft writing. Rough-draft talk and rough-draft writing means you are brainstorming, testing out ideas, making mistakes, making conjectures, and experimenting with numbers (Jansen, Cooper, Vascellaro, & Wandless, 2016). Nothing is final. Nothing gets erased. You will have one sheet of paper per strategy. Write the name of the strategy at the top. Solving 63 ÷ 9 may give you enough information to evaluate the strategy, or you may need to create additional problems. After working today, we will share our progress."

Since rough-draft talk and metacognition are emphases of today's lesson, Ms. Buchholz has purposefully paired students who speak the same heritage language. She hopes this will allow them the time and space to rough-draft talk in their heritage language before sharing with the whole class. Finally, Ms. Buchholz reminds students of the may-do task list posted on the whiteboard.

Scaffolding, Extending, and Assessing Student Thinking

As students work, Ms. Buchholz will perform three monitoring tasks. She will confer with pairs to hear their rough-draft talk about the must-do task. She will check in with students who move on to may-do tasks to hear their decision-making processes for selecting the task and what they are learning from it. In each conference, she will ask questions to scaffold and extend students' thinking. She will also have specific students complete a formative assessment called Show Me (Fennell, Kobett, & Wray, 2017) to check their understanding of previous learning intentions. Each of these monitoring tasks enables Ms. Buchholz to document her students' progress at varying levels of surface, deep, and transfer learning.

EFFECT SIZE FOR PROVIDING FORMATIVE EVALUATION = **0.48**

EFFECT SIZE FOR RECORD KEEPING = **0.52**

Ms. Buchholz uses two observation charts as she circulates the room. One chart has a space to record the task each student is working on and evidence of the success criteria (Figure 4.1).

The second chart has three variations of the Show Me task and the list of students Ms. Buchholz will meet with for each variation. The Show Me task is *Show Me how you would use _____ to represent the problem: There are 20 students and four tables. How many students should sit at each table to make four equal groups?* The variations are using cubes, pictures, or a number line to represent the division process. This formative assessment will allow Ms. Buchholz to assess students who were at the surface level of learning for the previous lesson's learning intention (*I am learning that the meaning of division is partitioning, sharing, and the inverse of multiplication*) but needed more time and experiences to progress to the deep level of learning.

Ms. Buchholz alternates between conferring with students and meeting with small groups for the Show Me tasks. This way, she has a clear, documented picture of what her students worked on independently as well as where her students are along the path toward mastery. She will use this documentation, student work, and students' sharing about their

Observation/Conference Chart

Dates: _____

Essential Questions: *What is multiplication? What is division? How are they related? How do we engage in the work of mathematicians?*

Adam Task:	Adoni Task:	Akeiyla Task:	Briera Task:	Chase Task:	Dillon Task:
Frances Task:	Geff Task:	Geneva Task:	Grace Task:	Hadley Task:	Kadisha Task:
Kayvion Task:	Leo Task:	Lobsang Task:	Machele Task:	Mario Task:	Monique Task:
Nathan Task:	Naum Task:	Pakmo Task:	Tahirah Task:	Ty Task:	Sabina Task:

Success Criteria	Questions
1. I can use and evaluate efficient division strategies. 2. I can use and evaluate efficient mathematical models to represent division strategies. 3. I can explain why a division strategy will always work using division language (*partitioning*, *sharing*, and the *inverse of multiplication*). 4. I can explain the inverse relationship between multiplication and division.	• Where is 62 in your representation? Where is 9? • What would be a story for $62 \div 9$? How is this story represented in your work? What does the answer mean? • How could I record numbers instead of drawing each object? • Will this strategy always work? Why or why not? • For what numbers would this strategy be inefficient? • How does this representation show why the strategy works?

Overall Patterns:

Figure 4.1

progress to plan tomorrow's math routines, mini-lesson, must-do and may-do tasks, and sharing.

Teaching for Clarity at the Close

As work time ends, Ms. Buchholz notes two patterns in her observation chart: only one pair of students moved on to may-do tasks and two students need additional practice using number lines to represent a contextualized division problem. Tomorrow, she will meet with these two students in a needs-based strategy group to model and practice using number lines.

On different days, the closing share has other focuses. Sometimes students share the content of their work, the craft or representations they choose, their processes or strategies, or their progress toward the learning intentions or a personal learning goal (Thunder & Demchak, 2012). Ms. Buchholz recognizes that tomorrow will be a continuation of today's work. The upcoming progress share (Figure 4.2) will be important for students as they record what they have completed and what they will do next.

The teacher says, "Today, we will bring our work to a close with a progress share. Everyone will share with a partner, a different partner than you worked with today. You will each share three reflections:

- What did you work on?
- How did this help you work toward the learning intentions and demonstrate the success criteria?
- What does this mean you need to work on tomorrow?

You will each have 3 minutes to talk. Decide who will be Partner A and who will be Partner B. Partner A, raise your hand. You go first."

Ms. Buchholz records the three reflection questions on the whiteboard and then listens in on her students' conversations. She makes additional notes in her observation chart. She gives a 30-second warning to wrap up and then announces Partner B's turn. She sees students looking back and forth between their work and the learning intentions and success criteria.

Teaching Takeaway

When measuring our impact on student learning, we must be prepared to make adjustments in the next steps of our teaching.

EFFECT SIZE FOR EVALUATION AND REFLECTION = 0.75

PROGRESS SHARE PROTOCOL

A Progress Share Protocol

Everyone shares with a partner, a different partner than you worked with today.

You will each share three reflections:

- What did you work on?

- How did this help you work toward the learning intentions and demonstrate the success criteria?

- What does this mean you need to work on tomorrow?

You will each have 3 minutes to talk. Decide who will be Partner A and who will be Partner B. Partner A begins.

Figure 4.2

She says, "At the beginning of class, there was some uncertainty and conjectures about the phrase *inverse of multiplication* in the learning intentions and success criteria. Based on your work today, what do you think this means?" Pairs join other pairs to have a small group discussion. Then they share ideas with the whole group. Students make connections like, "They're opposite operations," "You can think multiplication when you divide," and "It reminds me of addition and subtraction being inverses." Ms. Buchholz knows her students are beginning to make sense of this academic language through the problem-solving tasks and classroom discourse.

Next, Ms. Buchholz passes out a reflection paper with three sections. The first section is a box with the sentence starter: "Today I worked on." The second section is a table with each of the success criteria listed; each success criterion includes a dot to color green (got it), yellow (partially got it), or red (stuck/confused) and room to make a note with evidence of meeting or working toward that criterion. The third section is another box with the sentence starter: "Tomorrow I will." Ms. Buchholz says, "We will take 3 minutes to record reflections. When you evaluate your work on the success criteria, remember how we have evaluated ourselves as a class with colored sticky notes. This self-evaluation is like that but it only reflects *your* work and your thinking, rather than all of us." She observes, takes notes, and confers with students who have questions.

Ms. Buchholz initiates closure, saying "Place your reflection in your math binder. Decide who will store your rough-draft writing from today. Put away your mathematical toolbox; we will add a page of division strategies eventually. And then return to the carpet for quick images." When the majority of students have returned to the carpet, Ms. Buchholz begins flashing images of equal groups of dots. After each image, the students talk with a partner about how many dots they saw and how they saw them. This is a math routine that her students love. Ms. Buchholz knows the importance of subitizing for developing number sense, spatial reasoning, and mental math strategies (National Research Council, 2009). She also knows that spaced practice will help all of her students grow in their fluency while providing yet another context for making meaning of multiplication. The math class ends as it began, full of student talk and laughter. Figure 4.3 shows how Ms. Buchholz made her planning visible so that she could then provide an engaging and rigorous learning experience for her learners.

EFFECT SIZE FOR SPACED VS. MASS PRACTICE = 0.60

Ms. Buchholz's Teaching for Clarity PLANNING GUIDE

ESTABLISHING PURPOSE

1 **What are the key content standards I will focus on in this lesson?**

Indiana Academic Standards

> 3.C.5. Multiply and divide within 100 using strategies, such as the relationship between multiplication and division (e.g., knowing that $8 \times 5 = 40$, one knows $40 \div 5 = 8$), or properties of operations.

> 3.AT.2. Solve real-world problems involving whole number division within 100 in situations involving equal groups, arrays, and measurement quantities (e.g., by using drawings and equations with a symbol for the unknown number to represent the problem).

> 3.AT.5. Determine the unknown whole number in a division equation relating three whole numbers.

Standards for Mathematical Practice:

- Reason abstractly and quantitatively.

- Look for and express regularity in repeated reasoning.

2 **What are the learning intentions (the goal and *why* of learning stated in student-friendly language) I will focus on in this lesson?**

- Content: I am learning to evaluate fluent and nonfluent division strategies based on the representation and the numbers.

- Language: I am learning how the language of division (partitioning, sharing, and the inverse of multiplication) can be used to explain why a strategy will always work.

- Social: I am learning to reflect on the decisions we make as learners and how they help us grow as individuals and a community of learners.

3 When will I introduce and reinforce the learning intention(s) so that students understand it, see the relevance, connect it to previous learning, and can clearly communicate it themselves?

- Think-pair-share about learning intentions and success criteria
- Progress share and self-reflection

SUCCESS CRITERIA

4 What evidence shows that students have mastered the learning intention(s)? What criteria will I use?

I can statements:

- I can use and evaluate efficient division strategies.
- I can use and evaluate efficient mathematical models to represent division strategies.
- I can explain why a division strategy will always work using division language (partitioning, sharing, and the inverse of multiplication).
- I can explain the inverse relationship between multiplication and division.

5 How will I check students' understanding (assess learning) during instruction and make accommodations?

Formative Assessment Strategies:

- Conference/observation notes
- Show Me assessment
- Student work
- Progress self-reflection

Differentiation Strategies:

- Must-do and may-do tasks
- Purposeful pairing of students by heritage language

INSTRUCTION

6   What activities and tasks will move students forward in their learning?

- Oral counting math talk
- Must-do task: evaluating fluency
- May-do tasks: the Product Game, Multiplication Bingo, Multiples on Number Charts, Mysterious Multiplying Jar Problem, and More Division Situations
- Quick images

7   What resources (materials and sentence frames) are needed?

- Anchor chart of division strategies
- Word wall word: fluency
- Mathematical toolboxes
- Math binders
- Cubes
- Number charts
- Graph paper
- Open number lines and whiteboard markers
- Colored pencils
- Calculators
- Progress reflections
- Show Me task variations

8   How will I organize and facilitate the learning? What questions will I ask? How will I initiate closure?

Instructional Strategies:

- Must-do and may-do tasks

- Conferences
- Think-pair-share
- Rough-draft talk
- Self-reflection

Scaffolding Questions:

- Where is 62 in your representation? Where is 9?
- What would be a story for 62 ÷ 9? How is this story represented in your work? What does the answer mean?
- How could I record numbers instead of drawing each object?

Extending Questions:

- Will this strategy always work? Why or why not?
- For what numbers would this strategy be inefficient?
- How does this representation show why the strategy works?

Self-Reflection and Self-Evaluation Questions:

- What did you work on?
- How did this help you work toward the learning intentions and demonstrate the success criteria?
- What does this mean you need to work on tomorrow?
- What is your thinking now about the meaning of division as the "inverse of multiplication"?

 This lesson plan is available for download at **resources.corwin.com/vlmathematics-3-5**.

Figure 4.3   Ms. Buchholz's Procedural Knowledge Lesson on Fluent Division Strategies

Ms. Mills and Comparing Fractions

Ms. Mills and her class are in the midst of a deep dive into rational numbers. Throughout the last 2 weeks, they have been solving a variety of contextualized problems or fraction problem types (Empson & Levi, 2011), including equivalencing problems with price and rate contexts, comparison problems, and addition and subtraction problems with fractions with like denominators. The language of the classroom has shifted as students' academic vocabulary for talking and writing about fractions has steadily grown with explicit vocabulary instruction and a lot of opportunity to talk.

On the walls, there are several anchor charts with important fraction equivalencies. These serve as a way to track the classroom community's growing expertise with benchmark and equivalent fractions and is a reference for students while problem solving. The anchor charts and other student work around the room reveal concrete, representational, and abstract fraction models used for proofs, examples, and making connections. One type of concrete model are manipulatives. Ms. Mills knows the value of manipulatives is not *that* there are manipulatives available in her classroom, but *how* the manipulatives are used by the students to make their thinking and learning visible.

Ms. Mills wants her students to use their deep conceptual understanding of the unit big idea—*fractions are another way to represent division*—to develop procedural knowledge related to comparing and ordering fractions. She believes her students must build their procedural knowledge from conceptual understanding (NCTM, 2014) so that they can explain why a procedure works, evaluate whether a procedure will always work, and develop rules for procedures that will transfer to new situations. In today's lesson, the big idea is: *Fractions are equivalent (or equal) if they are the same size or the same point on a number line and they refer to the same size whole.* Her students will work with an assigned partner to compare fractions and then place all fractions on a number line. The fractions will be grouped to emphasize particular comparison strategies:

- Fractions with the same numerator,

- Fractions with the same denominator, and

- Comparing the distance of the fractions to benchmark fractions $\left(\frac{1}{2}, 1, 2\right)$.

EFFECT SIZE FOR
VOCABULARY
INSTRUCTION
= 0.62

Teaching Takeaway

Anchor charts support the development of assessment-capable learners by helping learners plan the next steps in their learning and select the right tools for learning.

EFFECT SIZE FOR
IMAGERY = 0.45

EFFECT SIZE FOR
MANIPULATIVE
MATERIALS ON
MATH = 0.30

EFFECT SIZE FOR
COGNITIVE TASK
ANALYSIS = 1.29

Video 12
Setting the Stage for
Procedural Learning

*https://resources.corwin.com/
vlmathematics-3-5*

The final step of the task will be to develop rules for comparing any fractions.

What Ms. Mills Wants Her Students to Learn

When we last peeked into Ms. Mills's classroom, her students were engaged in surface-level learning. They have moved along the continuum and are currently digging deeply into the procedures, strategies, and models related to fractions. Her lesson will address the following standards.

VIRGINIA MATHEMATICS STANDARD OF LEARNING

4.2. The student will (a) compare and order fractions and mixed numbers, with and without models, and (b) represent equivalent fractions.

Ms. Mills is helping her learners develop the following Mathematical Process Standards:

- Mathematical reasoning

- Mathematical representations

- Mathematical communication

These three mathematical ways of engaging with the content will enable her students to closely examine the structure of mathematics through the lens of fractions.

Learning Intentions and Success Criteria

Today's lesson marks a critical learning point in Ms. Mills's unit on fractions. While unit planning, Ms. Mills identified several places along students' learning paths where the learning intentions signaled significant procedural knowledge.

Content Learning Intention: I am learning to understand strategies for comparing fractions that rely on the value of fractions as numbers.

Language Learning Intention: I am learning to understand how fraction language, notation, and representations can be used to prove fraction comparisons and equivalencies.

Social Learning Intention: I am learning to understand that collaborating as a team means that mathematical disagreement and debate are valued and resolved respectfully.

By identifying these critical learning points prior to teaching the unit, Ms. Mills is able to plan exit tasks (Fennell et al., 2017) that check for mastery of critical procedural knowledge. She then analyzes student work on the exit tasks and adjusts her instruction to target her students' specific needs while within the deep learning phase of her unit.

> EFFECT SIZE FOR PROVIDING FORMATIVE EVALUATION = 0.48

Ms. Mills will begin her lesson by sharing the day's agenda. She plans to emphasize the learning intentions and success criteria that unite the agenda's activities. Each activity provides students with the opportunity to practice and make sense of the success criteria.

> EFFECT SIZE FOR RESPONSE TO INTERVENTION = 1.29

☐ I can compare fractions using strategies that rely on the value of fractions as numbers—the size of the piece, the number of pieces, and the distance to benchmark numbers.

☐ I can create a number line to show the comparisons, equivalencies, and density of fractions.

☐ I can use these comparisons to name additional fractions on the number line and to develop rules for comparing any fractions.

☐ I can prove, defend, revise, and refine my mathematical reasoning using fraction language, notation, and representations.

Ms. Mills wants her to students to see what mastery of the critical procedural knowledge looks like from the start of their work. During her mini-lesson, Ms. Mills will show students the exit task. She will think aloud (Trocki, Taylor, Starling, Sztajn, & Heck, 2014/2015) to model using the

> EFFECT SIZE FOR MASTERY LEARNING = 0.57

EFFECT SIZE FOR
EXPLICIT TEACHING
STRATEGIES = **0.57**

EFFECT SIZE FOR
SCAFFOLDING
= **0.82**

EFFECT SIZE
FOR TEACHING
COMMUNICATION
SKILLS AND
STRATEGIES = **0.43**

success criteria to reflect on her understanding of the learning intentions and to demonstrate mastery.

Activating Prior Knowledge

Math class follows a predictable structure in Ms. Mills's room: whole-group mini-lesson, work time, and whole-group sharing and closure. The math learning time begins or ends with one of several math routines, such as quick images, math talks, Which One Doesn't Belong?, and number talks. This predictable structure creates classroom cohesion and communicates consistent expectations to students. It reduces students' cognitive load so that they can focus on the rigorous mathematical content and processes within each lesson.

Today, Ms. Mills begins class by engaging in a math talk centered on oral counting: "We're counting by sixths today. As we count, I will record the count on a number line. Since we are counting by the same amount each time, I will do my best to keep the distance between each number the same. Ready. Begin."

The class counts chorally, "One-sixth, two-sixths, three-sixths, four-sixths . . . " as Ms. Mills creates a number line. At twelve-sixths, Ms. Mills says, "Stop. We counted from zero to twelve-sixths. Last week, several people pointed out that if you can count by ones, you can count by any fractional amount. You just have to say the number of pieces and the size of the piece. Where do you see this pattern in today's count?"

Benjamin explains, "Each time, we say the numerator, which is the number of pieces. This increased by one. And we also say the denominator or the size of the piece, which is always sixths."

"Benjamin is using our definitions of numerator and denominator to talk about the value of the fractions," Ms. Mills explains as she points to the word wall where students have created and posted cards with words, definitions, images, and examples. All of the words are related to fractions. Previous units' words are now on book rings available near the manipulative shelves for reference. When students discovered the usefulness of multiples and factors for problem solving with fractions, they retrieved those words from the multiplication and division unit and posted them back on the word wall.

Ms. Mills continues, "Find three fractions you can rename. Find one fraction you can rename in at least two ways. Turn and talk to a partner." This math routine allows her to give succinct directions, which maximizes her instructional time, because students know what to expect. Students list fractions with their partners. Some pairs make lists of equivalent fractions on whiteboards. Others look at anchor charts.

After 2 minutes, each pair calls out a pair of equivalent fractions while Ms. Mills adds to the number line. She asks, "Eight-sixths, one and two-sixths, one and one-third—how can all of these be equivalent?" Students take turns offering explanations, based on anchor charts, air drawings of fraction pieces, and contextualized scenarios.

Ms. Mills adds three more dashes to her number line after twelve-sixths and marks an "X" below the third dash. "What fraction would be here?" she asks. "How do you know? Turn and talk to a partner." After 1 minute, Ms. Mills asks for responses and lists them: fifteen-sixths, two and three-sixths, two and one-half, five-halves. When she has all responses (correct and incorrect), students take turns defending the numbers, which leads to students revising their responses, making connections between responses, and self-correcting.

> EFFECT SIZE FOR EVALUATION AND REFLECTION = 0.75

Ms. Mills's students are warmed up. They are reminding each other of fraction language, reorienting themselves to the resources around the room, and making connections to previous days' learning. Now, Ms. Mills introduces the agenda, followed by the learning intentions and success criteria, which are displayed on the whiteboard.

> Today, we are examining strategies for comparing fractions. All of the strategies must rely on the value of fractions as numbers, which means they have to make sense and be proven with fraction language, notation, and representations. We'll collaborate as a team of mathematicians. We know mathematicians value respectful, mathematical disagreement and debate. You'll know you've mastered today's learning intentions if you can compare fractions using strategies that rely on the value of fractions as numbers; develop rules for comparing any fractions; create a number line to show

comparisons, equivalencies, and additional fractions; and prove, defend, revise, and refine your reasoning using fraction language, notation, and representations. Turn and tell a partner what we are working on today and how we'll know we've got it.

EFFECT SIZE
FOR STRONG
CLASSROOM
COHESION = 0.44

Ms. Mills listens in on a few pairs and then continues.

At the end of our work time and sharing, I will give you an exit task.

MS. MILLS'S EXIT TASK ON EQUIVALENT FRACTIONS

Between which two fractions on our number line would you place _____?

Explain your reasoning. What strategy or rule did you use?

- I can compare fractions using strategies that rely on the value of fractions as numbers - the size of the piece, the number of pieces, the distance to benchmark numbers.

- I could use some help with using these strategies to compare fractions. I'm wondering _____.

- I'm stuck. I don't understand how to use one or more of the strategies to compare fractions.

I will fill in the number later. You'll solve the task and evaluate how close you are to mastering the learning intentions by choosing one response. Let's look back at our sixths number line and think about how our counting math talk could help us think about comparing fractions.

Ms. Mills circles two fractions: five-sixths and eight-sixths. She says, "Looking at this number line, I see five-sixths is less than eight-sixths.

How can we prove this using our number line? How can we prove this using fractional language for numerators and denominators?" Ms. Mills pauses to allow her students think time. Then she thinks aloud.

> I know the numerator tells how many pieces and because they are both sixths, the sizes of the pieces are all the same. So, eight-sixths is three more shares of sixths than five-sixths. That means eight-sixths is greater than five-sixths. I could also look at the numbers' distances to one whole or six-sixths. Five-sixths is one-sixth less than one whole and eight-sixths is two-sixths more than one whole. So five-sixths is less than eight-sixths.

As Ms. Mills talks aloud about her reasoning, she also points to the number line. Next, she instructs, "Turn and talk to a partner about the two strategies I used to prove that five-sixths is less than eight-sixths." She hears students talking about how to compare the distance to one whole and the number of pieces.

Then Ms. Mills returns to the learning intentions and success criteria.

> Based on this one problem, I fit the first category: *I can compare fractions using strategies that rely on the value of fractions as numbers*. I used the size of the pieces, the number of pieces, and the distance to the benchmark number one whole to compare the fractions. During our work time, you will work with an assigned partner to compare sets of fractions using strategies similar to the ones I modeled. Then you will order the fractions on a number line. Remember to represent the proportional distance between fractions as you place them on the number line. You and your partner will develop generalizable rules for comparing fractions. That means the rules have to always work for all fractions so you'll have to test them out, debate, and revise. I will ask people to share their rules. The last thing we'll work on is our exit task and self-evaluation.

Teaching Takeaway

The learning intentions and success criteria should serve as a guide for both the teacher and the learners. Referring back to them during instruction helps everyone know where they are going, how they are going, and where they are going next.

Ms. Mills answers clarifying questions and passes out the assignment sheets with partners' names listed. The assignment sheets have one of the following tiered or parallel tasks (Thunder, 2014):

- The low-readiness pairs compare four pairs of fractions, including fractions with the same numerator, fractions with the same denominator, fractions with a recognizable distance to one whole (comparing improper fractions or mixed numbers with fractions less than one), and fractions with a recognizable distance to one-half.

- The mid-readiness pairs compare pairs and trios of fractions, emphasizing the same four categories of comparing strategies.

- The high-readiness pairs also compare pairs and trios of fractions, emphasizing the same four categories of comparing strategies and a fifth category—a recognizable distance to two (comparing improper fractions and mixed numbers).

The students find their partners and a space to collaborate.

Scaffolding, Extending, and Assessing Student Thinking

During work time today, Ms. Mills plans to meet with three needs-based strategy groups and confer with partners to identify at least three who will share their fraction comparison rules. Ms. Mills used her previous days' conference notes and student work analyses to identify groups of two to five students with similar, targeted needs related to the unit's conceptual understandings and procedural knowledge. One group needs to revisit the reason for comparing fractions of the same-size whole. Another group needs to practice naming, renaming, and writing mixed numbers and improper fractions greater than one. A third group of students asked about halves of fractions; for example, they know four-eighths is equivalent to one-half, but they are wondering "What is four-and-a-half-ninths?"

Ms. Mills knows her students are at varying levels of surface, deep, and transfer learning and need deliberate practice with immediate, specific feedback. She also knows some students need additional time to explore a concept, learn notation, and practice academic language, whereas

EFFECT SIZE FOR "RIGHT" LEVEL OF CHALLENGE = 0.74

EFFECT SIZE FOR RECORD KEEPING = 0.52

EFFECT SIZE FOR DELIBERATE PRACTICE = 0.79

others are interested in pursuing concepts beyond the standards yet still significant to their quantitative reasoning. Meeting with flexible, needs-based strategy groups throughout the week allows Ms. Mills the opportunity to provide this deliberate practice in small groups for short periods of time while continuing to move the whole class forward.

> EFFECT SIZE FOR SPACED VS. MASS PRACTICE = **0.60**

Between each of her needs-based strategy groups, Ms. Mills circulates and confers with pairs. She values their engagement in productive struggle as problem solvers because she knows this is when they are learning. She chooses when to model academic language to describe fractional amounts, to ask focusing questions, and to refer students to the anchor charts for equivalencies. Ms. Mills encourages partners to think aloud while sketching in order to make their thinking visible to their peers. She knows peers' language, reasoning, and representations are often more powerful for students' sense-making than when the teacher does the majority of the talking and thinking. Ms. Mills planned focusing questions that reflect the learning intentions of the lesson, such as "Will this always work? What is another fraction between __ and __?" She has questions written on her conference note chart. Today, her chart is set up as a checklist to quickly note which pairs have developed which of the three types of rules for fractions, the targeted procedural knowledge of the lesson.

> EFFECT SIZE FOR RECIPROCAL TEACHING = **0.74**

Teaching for Clarity at the Close

The sound of the chime marks the end of work time, and Ms. Mills's students gather to share. Based on her conferences, Ms. Mills has recruited four pairs to share their strategies: one pair has a rule for comparing any fractions with the same denominator, one has a rule for comparing any fractions with the same numerator, one has a rule for comparing any fractions based on their distance to one-half, and the final pair wants to propose a rule for comparing any fractions based on their distance to any other common fraction.

As each pair shares their rule, they explain using two sample fractions. Then they place the fractions on a giant number line hanging across the wall and record additional equivalencies on anchor charts. Ms. Mills asks questions to make connections to the learning intentions:

> EFFECT SIZE FOR SUMMARIZATION = **0.79**

- How did they use the size of the pieces to compare?
- How did they use the number of pieces to compare?

Video 13

Direct/Deliberate
Instruction in a
Procedural Task

*https://resources.corwin.com/
vlmathematics-3-5*

EFFECT SIZE FOR
QUESTIONING
= 0.48

- How did they use the size of the missing piece or the distance to one whole to compare?

- How is their rule similar to or different from a rule you developed?

After the final pair shares, Ms. Mills's questioning shifts to highlight the number line model and the value of the fractions represented by length:

- How could we use the distance from zero to compare?

- How could we use the distance to any common fraction to compare?

- Why is comparing distance on the number line an accurate, efficient, and flexible strategy for comparing fractions?

To initiate closure, Ms. Mills's questioning facilitates students' transfer of these comparison strategies to making sense of the density of fractions:

- How can we use these rules to place any fraction on the number line? To find any fraction between two other fractions?

- Will there always be another fraction? Why or why not?

The students engage in rich discussion. Initially, there is much agreement; however, as the questions extend their thinking toward generalizability, strategy evaluation, and the density of fractions, the disagreement increases. Ms. Mills's progression of questions is purposeful. She wants the learning community to experience and value mathematical disagreement and debate. She is also comfortable with placing her students in discomfort as mathematical thinkers. To bring closure, Ms. Mills records students' burning questions, which she will use to plan her next conferences, needs-based strategy groups, and tasks, as well as their simmering questions, which she will use to plan extension and application tasks throughout the remainder of the unit.

Ms. Mills returns to the exit task, but this time she includes a fraction: *Between which two fractions on our number line would you place seven-tenths? Explain your reasoning. What strategy or rule did you use?* After completing the exit task, each student self-evaluates using the three indicators.

Ms. Mills wants to allow students to finish at varying times. As students turn in their exit task, they are invited to silently consider a problem posed through another typical math routine: Which One Doesn't Belong?

$$\begin{array}{|c|c|} \hline \dfrac{3}{4} & \dfrac{4}{6} \\ \hline \dfrac{2}{3} & \dfrac{5}{6} \\ \hline \end{array}$$

This task is both open middled and open ended, meaning there are multiple strategies for solving and multiple correct answers.

When all students have submitted their exit tasks and self-evaluations, Ms. Mills has students talk in triads about their solutions and reasoning. Then, she invites responses.

Gianna says, "Four-sixths doesn't belong because the other fractions are all one piece away from one whole."

Dashawn sees it differently and says, "Five-sixths doesn't belong because the other fractions are one piece away from one-half."

Luz relies on equivalent fractions and explains, "Three-fourths doesn't belong because the other fractions can all be named as sixths. Two-thirds can be four-sixths."

Ayla is quick to add, "I want to add on to what Luz said about three-fourths doesn't belong. The other fractions can all be named as thirds too. Four-sixths is two-thirds and five-sixths is two and a half-thirds." She was part of a needs-based strategy group working at the transfer level today.

Ms. Mills congratulates the class on their perseverance, use of mathematical language, and respectful debate. She and her students can see and hear their growth toward mastery of critical procedural knowledge within their unit on fractions. While they are at different levels of surface, deep, and transfer thinking, they have each taken significant steps forward in their learning today.

Figure 4.4 shows how Ms. Mills made her planning visible so that she could then provide an engaging and rigorous learning experience for her learners.

EFFECT SIZE FOR METACOGNITIVE STRATEGIES = 0.60

Video 14
Direct/Deliberate Instruction to Practice Mathematical Language and Precision

https://resources.corwin.com/ vlmathematics-3-5

Video 15
Consolidating Learning Through a Worked Example and Guided Practice

https://resources.corwin.com/ vlmathematics-3-5

Ms. Mills's Teaching for Clarity PLANNING GUIDE

ESTABLISHING PURPOSE

1 What are the key content standards I will focus on in this lesson?

Virginia Mathematics Standards of Learning

4.2 The student will (a) compare and order fractions and mixed numbers, with and without models, and (b) represent equivalent fractions.

Mathematical Process Standards:

- Mathematical reasoning
- Mathematical representations
- Mathematical communication

2 What are the learning intentions (the goal and *why* of learning stated in student-friendly language) I will focus on in this lesson?

- Content: I am learning to understand strategies for comparing fractions that rely on the value of fractions as numbers.
- Language: I am learning to understand how fraction language, notation, and representations can be used to prove fraction comparisons and equivalencies.
- Social: I am learning to understand that collaborating as a team means that mathematical disagreement and debate are valued and resolved respectfully.

3 When will I introduce and reinforce the learning intention(s) so that students understand it, see the relevance, connect it to previous learning, and can clearly communicate it themselves?

- Describe connections among the learning intentions, success criteria, and agenda before introducing the task
- Think aloud to model self-evaluation using success criteria

- Make connections during conferences and sharing
- Provide an exit task with self-evaluation

SUCCESS CRITERIA

4

What evidence shows that students have mastered the learning intention(s)? What criteria will I use?

I can statements:

- I can compare fractions using strategies that rely on the value of fractions as numbers—the size of the piece, the number of pieces, and the distance to benchmark numbers.
- I can create a number line to show the comparisons, equivalencies, and density of fractions.
- I can use these comparisons to name additional fractions on the number line and to develop rules for comparing any fractions.
- I can prove, defend, revise, and refine my mathematical reasoning using fraction language, notation, and representations.

5

How will I check students' understanding (assess learning) during instruction and make accommodations?

Formative Assessment Strategies:

- Conference notes checklist
- Student work
- Exit task with self-evaluation

Differentiation Strategies:

- Tiered or parallel tasks
- Purposeful pairing of students by readiness

INSTRUCTION

6 **What activities and tasks will move students forward in their learning?**

- Oral counting math talk
- Think-aloud
- Fraction comparisons tiered tasks
- Which One Doesn't Belong?

7 **What resources (materials and sentence frames) are needed?**

Anchor charts of equivalencies

Language frames

Large number line

Cuisenaire rods

Fraction pieces

Graph paper

Colored pencils

Whiteboards and markers

8 **How will I organize and facilitate the learning? What questions will I ask? How will I initiate closure?**

Instructional Strategies:

- Think-aloud with self-evaluation
- Needs-based strategy groups
- Anticipate, monitor, select, sequence, and connect students' comparison rules
- Turn and talk
- Reciprocal teaching/peer tutoring

Scaffolding Questions:

- How could you sketch what you are visualizing to show your partner your thinking?
- Which anchor chart could help you?

Extending Questions:

- Will this strategy always work?
- What is another fraction between ___ and ___? How do you know?

Connecting Questions:

- How did they use the size of the pieces to compare?
- How did they use the number of pieces to compare?
- How did they use the size of the missing piece or the distance to one whole to compare?
- How is their rule similar to or different from a rule you developed?
- Could we use the distance from zero to compare?
- How could we use the distance to any common fraction to compare?
- Why is comparing distance on the number line an accurate, efficient, and flexible strategy for comparing fractions?
- How can we use these rules to place any fraction on the number line? To find any fraction between two other fractions?
- Will there always be another fraction? Why or why not?

Self-Reflection and Self-Evaluation for Closure:

- Exit task with self-evaluation

 This lesson plan is available for download at **resources.corwin.com/vlmathematics-3-5**.

Figure 4.4   Ms. Mills's Procedural Knowledge Lesson on Comparing Fractions

Ms. Campbell and Computing Volume

After their initial experiences with volume and the idea of cubic units as an easy way to pack an object to measure volume, today's lesson focuses on developing deep understanding of the volume formula. Ms. Campbell wants her students to be able to use the formula with understanding; they should know what the terms mean and how to reason about volume, even if they do not remember the formula in the moment.

What Ms. Campbell Wants Her Students to Learn

In this part of the unit, Ms. Campbell is focused on developing procedural fluency. She expects her fifth grade students to be able to find the volume of right rectangular prisms with confidence by using the dimensions of the figure rather than packing a shape with cubes and then counting them. Her lesson today will set the stage for this fluency by helping students understand the meaning of the formula.

Because this lesson focuses on the volume formula, Ms. Campbell's lesson will focus on the standards related to developing and applying the formula for volume of a right rectangular prism: $V = l \times w \times h$. Students will work toward this standard by applying the mathematical practice of identifying and using structure.

Teaching Takeaway

Procedural fluency is more than just algorithms. Procedural fluency includes the matching of the right process with the right context—the meaning behind the process.

MATHEMATICS CONTENT AND PRACTICE STANDARDS

5.MD.C. Understand concepts of volume and relate volume to multiplication and to addition.

5. Relate volume to the operations of multiplication and addition and solve real-world and mathematical problems involving volume.

a.   Find the volume of a right rectangular prism with whole-number side lengths by packing it with unit cubes and show that the volume is the same as would be found by multiplying the edge lengths, equivalently by multiplying the height by the area of the base. Represent threefold whole number products as volumes, e.g., to represent the associative property of multiplication.

b. Apply the formula $V = l \times w \times h$ and $V = b \times h$ for rectangular prisms to find volumes of right rectangular prisms with whole number edge lengths in the context of solving real-world and mathematical problems.

Ms. Campbell is helping her learners develop the following Standards for Mathematical Practice:

• Look for and make use of structure.

Ms. Campbell has struggled herself to understand this idea of structure in mathematics. Based on her own developing understanding, she knows that structure is about underlying relationships or organizing principles. In this case, the formula for volume represents the structure of right rectangular prisms. By making the connection between the two forms for the formula ($V = l \times w \times h$ and $V = b \times h$), the stage is set for finding the volume of other shapes where we know the area of the base and the height. Although students will not extend their learning this far in fifth grade, Ms. Campbell has come to understand that she teaches more effectively when she knows where student learning is headed in higher grades. In Grade 8, students will use this idea to find the volume of cylinders. Although it feels like one more detail today, this broader perspective helps Ms. Campbell help her students to understand both forms of the formula.

Learning Intentions and Success Criteria

While the standards anchor Ms. Campbell's thinking about the content of the unit, daily instruction in each lesson is grounded in clear learning intentions and success criteria. The following specific learning intentions help her students know what to focus on today.

> EFFECT SIZE FOR PROFESSIONAL DEVELOPMENT PROGRAMS = 0.41

Teaching Takeaway

Teaching mathematics in the Visible Learning classroom requires teachers to see learning as part of a progression, aware of what is coming in future years of study.

> EFFECT SIZE FOR COGNITIVE TASK ANALYSIS = 1.29

Content Learning Intention: I am learning how to compute the volume of right rectangular prisms based on edge lengths and the area of one face.

Language Learning Intention: I am learning the connection between the vocabulary of solid figures (*edge, face*) to the formats for the volume formula ($l \times w \times h$ or $b \times h$)

Social Learning Intention: I am learning how to explain my thinking clearly to my partners.

The following success criteria will show her students when they are on the right track in meeting the learning intentions for the day.

☐ I can compute the volume of a right rectangular prism using the formula $l \times w \times h$.

☐ I can compute the volume of a right rectangular prism using the formula $b \times h$.

☐ I can explain why each formula always works for right rectangular prisms.

☐ I can use the measurement of side lengths to help me compute the volume of a figure efficiently.

Activating Prior Knowledge

Ms. Campbell begins the lesson by reminding students of their findings from earlier lessons.

> Class, one of the things we have noticed in our work with volume so far is that it is hard to count the unit cubes when they are inside a box that is not clear. We have also noticed that our boxes do not always fit our cubes perfectly. It's time to get more precise. For today's lesson, we are going to be building right rectangular prisms rather than packing the cubes into boxes to measure volume. We will be finding the volume of the shape we build rather than measuring the inside of the box.

> Today we will use dice to determine how big to make our
> prisms. Each pair of you will have three dice because our
> prisms have three dimensions—how wide they are, how long
> they are, and how high they are.

As she says this, Ms. Campbell picks up one of the boxes the class has used for earlier activities and points to each dimension as she names it.

She continues, "Our learning intentions for today are about using the formula for volume, including good vocabulary, and explaining our thinking clearly." Ms. Campbell briefly talks through each learning intention before asking students to talk with a shoulder partner.

EFFECT SIZE FOR
STRATEGY TO
INTEGRATE WITH
PRIOR KNOWLEDGE
= 0.93

> Describe the learning intentions to your partner in your own
> words. Make sure you both understand what our purpose
> is today. After this, read the success criteria and be ready to
> ask any questions you have about how we will know we are
> learning the right mathematics today.

EFFECT SIZE FOR
PLANNING AND
PREDICTION = 0.76

Ms. Campbell gives the class time for a short conversation as she listens to the students talk. She hears many students struggling to read the formulas, naming the multiplication symbol the letter "X," and not knowing what the letters stand for.

EFFECT SIZE
FOR PROVIDING
FORMATIVE
EVALUATION = 0.48

"Conceda, would you share your observation about the formulas with the class?" Ms. Campbell asks. "You recognized part of the formula as something you've seen before."

Conceda says, "I saw length times width and thought about area. That's the formula we use for area and here it is again."

"I don't see the words *length*, *width*, or *times* in our lists. What are you seeing there that tells you length times width?" the teacher asks.

"The formula for area is $l \times w$, and that means length times width. We multiply the sides of a rectangle to find the area," Conceda explains.

The teacher continues, "I heard some students struggling to read the formulas while you were talking in your pairs. Go back now that you've heard Conceda's explanation and practice reading the formulas for volume. The letter b stands for base. You'll learn what that means during the lesson."

Teaching Takeaway

Providing formative evaluation allows teachers and learners to make their learning visible and better map out the next steps in the learning journey. Formative evaluation informs our judgment about where to go next.

Teaching Takeaway

Collaborating around student learning is one of the key mindframes for teaching mathematics in the Visible Learning classroom.

Ms. Campbell is pleased she noticed Conceda's observation about area as she listened. Because Ms. Campbell has only taught fifth grade, she has realized limits in her understanding of the math her students are learning. Over the past few years, her personal professional growth goal has been to develop a broader understanding of the math her colleagues teach before and after fifth grade. Her principal has supported this work by giving her time to observe in her colleague's classrooms, and she has taken her own time to talk with them outside of school when she has questions. She visited her third grade colleague during their area unit and this helped her make this connection.

Scaffolding, Extending, and Assessing Student Thinking

Ms. Campbell says, "Let me show you today's activity. Each pair of students will need a recording sheet and three dice." She has placed recording sheets on each table and distributes bags of three dice to each pair. She is differentiating the content of this lesson by the readiness of the students for noticing patterns in multiplication. Most students have three 6-sided dice, while a few have an 8- or 12-sided die added in. For one pair still struggling with multiplication fluency, Ms. Campbell has three 4-sided dice.

Recording Sheet for Computing Volume

Figure	Length	Width	Height	Volume (Number of Cubes)	Area of Base	Height
1						
2						
3						
4						

Ms. Campbell continues the lesson.

> Let's look at the recording sheet first. What columns do
> you see? You'll build at least four figures during your work
> time. There's space to record more if you have time. For each
> figure, roll your three dice and decide which number will be
> the length, which one the width, and which one the height.
> When you roll, remember that if you have a four-sided die,
> the roll is the number repeated around the base of the die, not
> the number on top.

She deliberately gives this instruction to the whole class so she does
not highlight which students have the four-sided dice. She will confirm
their understanding when the class starts working.

> Once you know the dimensions, use your cubes to build your
> prism. Then find the volume of the prism using the strategy
> of your choice. In the last two columns, you will look at your
> figure a different way. The base of the figure is the side that is
> sitting on the table. It's the same shape as the top of the figure.
> Find the area of the base and then record how tall the shape is.

> As you build your figures and record information, look for
> patterns you might use to speed up your work. Remember, our
> fourth success criterion is about efficiency. While counting
> will get you to the right answer, it can take a long time for
> big shapes. In this lesson, I want you to think about a more
> efficient way to find volume than counting.

Judy, Alex, and Nancy are working together with three 6-sided dice.
Their second roll is 2, 3, 5. "This means we have to build a rectangle
with six units and then make five layers," Alex observes.

"Wait a minute," says Judy. "We didn't roll a six. Where do you see
a six?"

Alex responds, "See the two and the three? If I build a rectangle with six
cubes, it has two rows of three or six cubes. That's length and width. The
other number is the height, so we need five rows of six."

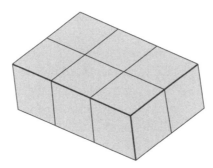

Nancy picks up the recording chart and starts to fill in the numbers on the second row.

Figure	Length	Width	Height	Volume (Number of Cubes)	Area of Base	Height
1	4	2	3	24	8	3
2	2	3	5	30	6	5
3						
4						

She records length, width, and height as Alex described and asks her group members to count how many cubes there are. Alex starts to add 6 + 6 + 6 . . . when Judy jumps in and says, "I see! You can multiply six times five. There are six in each row and that's like five groups of six. The volume is 30." Nancy writes 30 in the chart and asks her group what "area of the base" means.

Judy tells her that the area of the base is "the area of the part sitting on the table. That's the base. That's why Alex said we could multiply six by five to get the volume. That's the last part of the table. Now I understand the eight on the first row."

Ms. Campbell comes by their table and asks about their progress. Alex says, "We think we figured out that the base is the area of the bottom layer, and that's the same as the length times the width. That matches the formulas you have in our success criteria." After Alex summarizes their thinking, Ms. Campbell responds, "Can you test this conjecture with your next roll? Can you predict the volume before you build the shape?"

As Ms. Campbell observes and talks with other groups, she notices some groups are moving smoothly, including the team of Alex, Nancy, and Judy. Other groups are struggling. Because it is still early in the year, Ms. Campbell is careful to note which students need support on the mathematics and which students need help learning to work together. Both lessons are important when students are learning collaboratively. She can build additional strategies for working together into future lesson plans based on the struggles she has seen today.

> EFFECT SIZE FOR SCAFFOLDING = 0.82

Teaching for Clarity at the Close

After each team has built and calculated volume for at least three figures, Ms. Campbell brings the class together. "Let's review our success criteria for today," she says. "The first two items are about using the formulas for volume. Take a moment and self-assess your progress there. If you know how many cubes long each side of the prism is, can you find the volume both ways?"

> EFFECT SIZE FOR EVALUATION AND REFLECTION = 0.75

Ms. Campbell asks, "Who would like to share their thinking about why the formulas always work for this shape? Conceda?"

Conceda replies, "I think the two formulas say the same thing in different ways. If you multiply length times width, you get the area of the base. If you know how big the base is, you just have to count up the layers."

The teacher says, "Thank you, Conceda. I like the way you explained that the two formulas are related. Would anyone like to add on to Conceda's thinking? Judy?"

Judy says, "The number of cubes is the length of each part of the prism. This is like when we learned about area in third grade, except now we have height too. The area formula works for every rectangle and now this one works for every prism like this."

Ms. Campbell has found it helps to have students model the kind of thinking she wants to hear and to remind students that they do not have to master each learning intention in a single day.

> Class, turn to your partner and explain in your own words why the formulas work for finding the volume of prisms. Conceda and Judy have given us some good thinking to start with. After you explain, self-assess your progress. We'll use these ideas again tomorrow so you will have time to work on this if you aren't there yet.

> We have one last success criterion to talk about: *I can use the measurement of side lengths to help me compute the volume of a figure efficiently.* Could you find the volume of a box if I told you the length of the sides? Would you need the cubes? Turn to your partner and answer this question together in your journal.

She holds up a box and gives the class the measurements:

> This box is 12" wide, 10" long, and 6" high. What is the volume of the box?

As the class works, Ms. Campbell realizes she is out of time for this class. She had hoped to discuss this question today but will hold the discussion until tomorrow. She says, "Class, thank you for your good work today. We will return to this problem tomorrow. Please take the last minutes of class to record your thinking and return your cubes to their bag." Today's homework, practice computing the volume of boxes with given dimensions, will be delayed until tomorrow when students are ready.

Figure 4.5 shows how Ms. Campbell made her planning visible so that she could then provide an engaging and rigorous learning experience for her learners.

EFFECT SIZE FOR ELABORATION AND ORGANIZATION = 0.75

EFFECT SIZE FOR HOMEWORK = 0.29

Teaching Takeaway

For homework to be most effective in student learning, homework must practice something that has already been taught and that the learner can achieve success on his or her own. Homework where learners may struggle is not helpful.

Ms. Campbell's Teaching for Clarity PLANNING GUIDE

ESTABLISHING PURPOSE

1 What are the key content standards I will focus on in this lesson?

Content Standards:

> *5.MD.C. Understand concepts of volume and relate volume to multiplication and to addition.*

> *5. Relate volume to the operations of multiplication and addition and solve real-world and mathematical problems involving volume.*

> *a. Find the volume of a right rectangular prism with whole-number side lengths by packing it with unit cubes and show that the volume is the same as would be found by multiplying the edge lengths, equivalently by multiplying the height by the area of the base. Represent threefold whole number products as volumes, e.g., to represent the associative property of multiplication.*

> *b. Apply the formula $V = l \times w \times h$ and $V = b \times h$ for rectangular prisms to find volumes of right rectangular prisms with whole number edge lengths in the context of solving real-world and mathematical problems.*

Standards for Mathematical Practice:

- *Look for and make use of structure.*

2 What are the learning intentions (the goal and *why* of learning stated in student-friendly language) I will focus on in this lesson?

- *Content: I am learning how to compute the volume of right rectangular prisms based on edge lengths and the area of one face.*

- *Language: I am learning the connection between the vocabulary of solid figures (edge, face) to the formats for the volume formula ($l \times w \times h$ or $b \times h$)*
- *Social: I am learning how to explain my thinking clearly to my partners.*

3 When will I introduce and reinforce the learning intention(s) so that students understand it, see the relevance, connect it to previous learning, and can clearly communicate it themselves?

- *Introduce it at the beginning of the lesson so students have a clear map for their progress today.*

SUCCESS CRITERIA

4 What evidence shows that students have mastered the learning intention(s)? What criteria will I use?

I can statements:

- *I can compute the volume of a right rectangular prism using the formula $l \times w \times h$.*
- *I can compute the volume of a right rectangular prism using the formula $b \times h$.*
- *I can explain why each formula always works for right rectangular prisms.*
- *I can use the measurement of side lengths to help me compute the volume of a figure efficiently.*

5 How will I check students' understanding (assess learning) during instruction and make accommodations?

Formative Assessment Strategies:

- *Conference/observation notes*
- *Show Me*
- *Student work*

Differentiation Strategies:

- *Content by readiness: dice used for rolling prism sides–mix of four-, six-, and eight-sided dice depending on the fact fluency of the students.*

INSTRUCTION

6   **What activities and tasks will move students forward in their learning?**

- *Building Prisms activity*
- *Search for a Pattern*

7   **What resources (materials and sentence frames) are needed?**

- *Four-, six-, and eight-sided dice*
- *Unit cubes (in centimeters)*
- *Recording sheet*

8   **How will I organize and facilitate the learning? What questions will I ask? How will I initiate closure?**

Instructional Strategies:

- *Partner work*
- *Modeling with cubes*
- *Dice to determine dimensions of solid figures*

Scaffolding Questions:

- *What are the dimensions of your prism?*
- *Show me each dimension in your model.*
- *Where is the base of your prism?*

Extending Questions:

- *Does it matter which face is the base of the prism?*

- *Would your answer change if you built your prism with a different base?*

- *Can you predict the volume before you build the shape?*

- *Is there a way to find the volume without counting?*

Self-Reflection and Self-Evaluation Questions:

- *How did this lesson help you work toward the learning intentions and demonstrate the success criteria?*

- *What do you need to practice more or understand better?*

online resources ↗ | This lesson plan is available for download at **resources.corwin.com/vlmathematics-3-5**.

Figure 4.5   Ms. Campbell's Procedural Knowledge Lesson on Computing Volume

Reflection

Our final visit to these three classrooms focused on the development of procedural knowledge and fluency. Using what you have read in this chapter, reflect on the following questions:

1. In your own words, describe what teaching for procedural knowledge looks like in your mathematics classroom.

2. How does the Teaching for Clarity Planning Guide support your intentionality in teaching for procedural knowledge?

3. Compare and contrast the approaches to teaching taken by the classroom teachers featured in this chapter.

4. Consider the following statement: *Procedural knowledge is more than "drill and kill."* Do you agree or disagree with the statement? Why or why not? How is this statement reflected in this chapter?

5. How did the classroom teachers featured in this chapter adjust the difficulty and/or complexity of the mathematics tasks to meet the needs of all learners?

KNOWING YOUR IMPACT: EVALUATING FOR MASTERY

5

CHAPTER 5 SUCCESS CRITERIA:

(1) I can describe what mastery learning is in my classroom.

(2) I can compare and contrast checks for understanding with the evaluation of mastery.

(3) I can explain how to evaluate mastery in my own classroom using tasks and tests.

(4) I can identify characteristics of challenging mathematics tasks.

(5) I can explain the role of feedback in supporting students' journey to mastery.

EFFECT SIZE
FOR PROVIDING
FORMATIVE
EVALUATION = 0.48

EFFECT SIZE FOR
FEEDBACK = 0.70

Mastery learning is the expectation that learners will grasp specific conceptual understanding, procedural knowledge, and the application of specific concepts and thinking skills.

EFFECT SIZE FOR
MASTERY LEARNING
= 0.57

Teaching Takeaway

Effective feedback is an essential feature of the Visible Learning mathematics classroom.

EFFECT SIZE FOR
TEACHER CLARITY
= 0.75

Let us end right where we began—Ms. Showker's fourth grade classroom. Ms. Showker established clear learning intentions and success criteria, and she designed a challenging mathematics task that allowed learners to see themselves as their own teachers. Just like Ms. Buchholz, Ms. Mills, and Ms. Campbell, Ms. Showker created many opportunities for learners to make their thinking visible through her checks for understanding. Formative evaluation and feedback are critical components to teaching mathematics in the Visible Learning classroom.

However, this chapter focuses on determining students' learning over the long haul. In other words, how do teachers assess for mastery? And in doing so, how do teachers and learners make evidence-informed decisions about when to move forward in the learning progression? Knowing our impact on student learning in mathematics involves more than just formative evaluation of learning. Knowing our impact also involves recognizing student mastery in their mathematics learning.

What Is Mastery Learning?

Mastery learning is the expectation that learners will grasp specific conceptual understanding, procedural knowledge, and the application of specific concepts and thinking skills. This requires that teachers establish clarity about the learning in mathematics classrooms and then organize a series of logical experiences, noticing which students do and don't learn along the way. When students experience lesson clarity, they progress toward mastery. The claim underlying mastery learning is that all children can learn when provided with clear explanations of what it means to "master" the material being taught. Although mastery learning does not speak to the time learners need to reach mastery, all students continuously receive evaluative feedback on the performance. Learners know where they are at in their learning, where they are going, and what they can do to bridge the gap.

In true mastery learning, students do not progress to the next unit until they have mastered the previous one. But "moving on" could mean that learners move forward in the learning progression or that they are provided additional learning experiences at the surface, deep, or transfer level to address gaps in their learning if they are not yet able to demonstrate mastery. Ms. Showker notes,

To evaluate student mastery, I develop a rubric that describes the learning progression from the very basic level of learning all the way up to mastery—for example, Levels 1 through 4, with 4 being mastery. My plan for each student or group of students depends on the specific descriptions of each level and where that student is in terms of level. When learners demonstrate, say, a Level 1 or Level 2, I use this information to provide additional scaffolding for these learners. They are not there, yet.

Mastery learning is an essential part of building assessment-capable visible learners in the mathematics classroom. If learners are to know where they are going next in their learning, select the right learning tools to support the next steps (e.g., manipulatives, problem-solving approaches, and/or metacognitive strategies), and know what feedback to seek about their own learning, they must have opportunities to assess their own mastery with mathematics content. This, of course, comes after learners have engaged in multiple mathematics tasks replete with checks for understanding that allow teachers and students to adjust learning in the moment. Once that has occurred, it is time to determine students' level of mastery in the mathematics learning. So how do we determine what mastery looks like for specific content in the mathematics classroom?

Using Learning Intentions to Define *Mastery Learning*

Learning intentions provide the framework for defining *mastery* in learning, developing the assessments used to determine student mastery, and gathering the information necessary to plan learning experiences for students. Ms. Showker, Ms. Buchholz, Ms. Mills, and Ms. Campbell had to answer the question "What do my students need to learn?" The answer to this question represents mastery for the specific content in each of their classrooms. In the above example from Ms. Showker's classroom, the learning intention stated *I am learning that the type of data and the way I display that data are connected.* Therefore, to demonstrate mastery, her learners must organize and represent data in graphs. For this specific standard(s), her learners only need to demonstrate mastery

Teaching Takeaway

Features of mastery learning include the following:

1. Clear learning expectations

2. Feedback that is specific, constructive, and timely

3. Sufficient time, attention, and support to ensure learning

in representing data in line graphs and bar graphs (Virginia Department of Education, 2016).

Assessments of mastery require both the teacher and the learners to focus on the essential learning for a particular unit or series of lessons. Teachers must unpack the language of the specific standard to have a clear sense of the conceptual understanding, procedural knowledge, and applications expected in the mastery of the particular standard. Let us look at the specific content standard associated with Ms. Showker's lesson in Chapter 1. How would she assess for mastery?

The student will:

    a. collect, organize, and represent data in bar graphs and line graphs;

    b. interpret data represented in bar graphs and line graphs; and

    c. compare two different representations of the same data (e.g., a set of data displayed on a chart and a bar graph, a chart and a line graph, or a pictograph and a bar graph).

As you can see, this is not very helpful in developing and implementing an assessment of mastery learning. For example, how much data are her fourth graders expected to collect? Are the questions or investigations that produce the data teacher generated, student generated, or both? At what level must her learners interpret the data, and are her learners encouraged to use technology within this standard (e.g., graph by hand or use an electronic device)? Looking closer at the curriculum documents that accompany this specific standard, Ms. Showker is able to narrow in on tasks that will allow students to demonstrate mastery. Learners are only focusing on line graphs and bar graphs in fourth grade. To effectively evaluate for mastery, teachers must specifically define what the learner will know, understand, and be able to do. Ms. Showker and her collaborative planning team defined *mastery* as follows:

> Given an authentic scenario, I can formulate questions
> necessary for completing the task, collect and organize the

data generated from those questions, represent the data in the most appropriate type of graph (line graph or bar graph), interpret the findings to address the specific task, and write a mathematics story that explains the process for completing the task.

The collaborative planning team came up with the following scenarios from which learners could select (adapted from www.definedstem .com):

1. Memorial Elementary is changing the mascot of our school. Your task is to survey students throughout the school so that you can make a recommendation about the type of mascot and his or her specific characteristics (e.g., type of animal, color, and name).

2. The local fast-food restaurant is trying to offer healthier side options in the "kid's meals". Your task is to poll students in the school so that you can make a recommendation about side options that are both appealing to your peers and healthy (e.g., types of fruits, types of veggies, mixture of the two, etc.).

3. The principal is creating a safety manual for the school that will help prepare teachers and students for different types of severe weather or natural disasters. Your task is to provide data that help her decide how to prioritize those events in the safety manual (e.g., which one should be listed first, which one is not likely to occur, etc.).

In addition, Ms. Showker and her colleagues have specifically identified vocabulary words that represent key concepts within this standard that learners must use fluently in their work: *surveys, polls, questionnaires, scientific experiments, observations, axes, horizontal, vertical, categories, key, analysis, interpretation*, and *prediction*.

In Figure 5.1, Ms. Buchholz, Ms. Mills, and Ms. Campbell also define mastery for their particular standard(s) or learning intention.

EXAMPLES OF MASTERY FOR SPECIFIC CONTENT STANDARDS

	Mathematics Content Standard	What Mastery Looks Like
Ms. Buchholz	**From Chapter 3:** Represent the concept of multiplication of whole numbers with the following models: equal-sized groups, arrays, area models, and equal "jumps" on a number line. Understand the properties of 0 and 1 in multiplication. Solve real-world problems involving whole number multiplication within 100 in situations involving equal groups, arrays, and measurement quantities (e.g., by using drawings and equations with a symbol for the unknown number to represent the problem). Interpret a multiplication equation as equal groups (e.g., interpret 5 × 7 as the total number of objects in 5 groups of 7 objects each). Represent verbal statements of equal groups as multiplication equations. Create, extend, and give an appropriate rule for number patterns using multiplication within 100.	Given an authentic scenario (e.g., books on bookshelves, drinks in a package), learners must model the scenario using drawings and equations with a symbol for the unknown number to represent the problem, find a solution, and interpret the mathematical results in the context of the scenario.
Ms. Mills	**From Chapter 4:** The student will (a) compare and order fractions and mixed numbers, with and without models, and (b) represent equivalent fractions.	Given an authentic scenario (e.g., comparing unit prices of packaged items or rates of activity by athletes), learners must model the scenario using equivalent fractions, calculate a solution, and interpret the mathematical results in the context of the scenario.
Ms. Campbell	**From Chapter 2:** Understand concepts of volume and relate volume to multiplication and to addition. Relate volume to the operations of multiplication and addition and solve real-world and mathematical problems involving volume. Recognize volume as additive. Find volumes of solid figures composed of two non-overlapping right rectangular prisms by adding the volumes of the non-overlapping parts, applying this technique to solve real-world problems.	Given a choice of several composite shapes, learners must find the volume, compare the volumes of several composite shapes, and find unknown dimensions given a volume and one dimension.

Figure 5.1

Establishing the Expected Level of Mastery

From the preestablished levels of mastery, which is based on the standard(s), teachers can identify measurable indicators that students are or are not at the level of mastery. These indicators should focus on what students are doing rather than what they are not doing. This helps identify current performance levels and is suggestive of the types of experiences students need to have to progress in their learning. In other words, what does progress toward mastery look like in this specific standard? Learners progress toward mastery at different rates, and teachers should map out that progress so that both the teacher and the learners can make an informed decision about where they are in their learning.

Ms. Showker and her colleagues identified the incremental steps along the pathway to achieving mastery for the functional relationship standard using the SOLO Taxonomy (see Figure 1.5 in Chapter 1). If students perceive or actually are far from meeting the highest level of proficiency, making the progression visible allows them to answer the questions "Where am I going, how am I going, and where to go next?" These are essential in developing assessment-capable visible learners. Together, Ms. Showker and her colleagues developed the progression shown in Figure 5.2.

Teaching Takeaway

In addition to knowing what we want our students to learn, we have to know what evidence will demonstrate that they have learned it.

Video 16
Setting the Stage for Transfer

https://resources.corwin.com/vlmathematics-3-5

EXAMPLE OF PROGRESS TOWARD MASTERY FOR A SPECIFIC CONTENT STANDARD

	Unit: Probability and Statistics					
Content Standard The student will: (a) collect, organize, and represent data in bar graphs and line graphs; (b) interpret data represented in bar graphs and line graphs; and (c) compare two different representations of the same data (e.g., a set of data displayed on a chart and a bar graph, a chart and a line graph, or a pictograph and a bar graph).	**Learning Intention:** I am learning that the type of data and the way I display that data are connected.					**Vocabulary** surveys, polls, questionnaires, scientific experiments, observations, axes, horizontal, vertical, categories, key, analysis, interpretation, prediction
	How will I know when I have it? The following mastery levels will let you know how you are progressing toward this learning goal.					
	Level 4	*Level 3*	*Level 2*	*Level 1*	*Level 0*	**Prior Knowledge** skip counting
	I was able to formulate questions necessary for completing the task, collect and organize the data generated from those questions, represent the data in the most appropriate type of graph, interpret the findings to address the specific task, and write a mathematics story that explains the process for completing the task.	I was able to formulate questions specific to the task and organize the data. I included the essential characteristics of a graph. I could explain why I decided to represent the data with a specific type of graph. My mathematics story was focused on the process for completing the task. However, I struggled to make a clear recommendation based on my data.	I was able to formulate questions specific to the task and organize the data. I struggled to include the essential characteristics of a graph. My mathematics story was focused on the scenario and not on my process for completing the task. I described the scenario and did not interpret my data.	I was able to formulate questions specific to the task, but I struggled to organize the data and represent the data in a graph. I was not able to interpret my findings or write a mathematics story that explained my process.	I am not able to formulate questions, collect and organize data, represent the data, interpret my findings, and write a mathematics story explaining my process.	

Source: Adapted from Ashley Norris, Mathematics Teacher, Columbia County Public Schools, Georgia.

Figure 5.2

When we revisit the classrooms of Ms. Buchholz, Ms. Mills, and Ms. Campbell, we see that they provide similar levels of clarity about what mastery looks like as their learners progress through the big ideas around the content standards (see Figure 5.3). Teachers know their students best and therefore can use evaluation of student learning—within a set of learning

EXAMPLES OF LEVELS OF PROFICIENCY FOR SPECIFIC CONTENT STANDARDS

	Mathematics Content Standard	Levels of Proficiency—Progress Toward Mastery
Ms. Buchholz	**From Chapter 3:** Represent the concept of multiplication of whole numbers with the following models: equal-sized groups, arrays, area models, and equal "jumps" on a number line. Understand the properties of 0 and 1 in multiplication. Solve real-world problems involving whole number multiplication within 100 in situations involving equal groups, arrays, and measurement quantities (e.g., by using drawings and equations with a symbol for the unknown number to represent the problem). Interpret a multiplication equation as equal groups (e.g., interpret 5×7 as the total number of objects in 5 groups of 7 objects each). Represent verbal statements of equal groups as multiplication equations. Create, extend, and give an appropriate rule for number patterns using multiplication within 100.	**Level 1:** Shows minimal attempt on the problem (guess and check); provides no clear modeling or problem-solving approach (does not use drawings, equations, or symbols); provides no interpretation of the mathematical results; or provides no answer. **Level 2:** Shows signs of coherent problem solving through an attempt to model; gives minimal evidence to support the answer; fails to address some of the constraints of the problem; occasionally makes sense of quantities in relationships in the problem; has trouble generalizing or interpreting the mathematical results. **Level 3:** Response shows the main elements of solving the problem and an organized approach to solving the problem; models the problem using drawings, equations, or symbols—there are errors, but of a kind that the student could well fix, with more time for checking and revision and some limited help; makes sense of quantities and their relationships in the specific situation; response uses interpretation of mathematical results. **Level 4:** Shows understanding and use of stated assumptions, definitions, and previously established mathematical concepts in construction of models; makes conjectures and builds a logical progression of statements in the authentic scenario; routinely interprets mathematical results in the context of the situation and reflects on whether the results make sense; communication is precise, using definitions clearly.
Ms. Mills	**From Chapter 4:** The student will (a) compare and order fractions and mixed numbers, with and without models, and (b) represent equivalent fractions.	**Level 1:** Shows minimal attempt on the problem (guess and check); has no clear problem-solving approach that utilizes equivalent fractions; provides no reasoning with the answer or the interpretation of mathematical results; or provides no answer. **Level 2:** Shows signs of coherent problem solving and the use of equivalent fractions; gives minimal mathematical evidence to support the answer; fails to address some of the constraints of the problem; occasionally makes sense of quantities in relationships in the problem; has trouble generalizing, interpreting, or using the mathematical results. **Level 3:** Response shows the main elements of solving the problem and an organized approach to solving the problem using equivalent fractions; there are errors, but

(Continued)

	Mathematics Content Standard	Levels of Proficiency—Progress Toward Mastery
		of a kind that the student could well fix, with more time for checking and revision and some limited help; makes sense of quantities and their relationships in the specific situation; response uses assumptions, definitions, and previously established concepts.
		Level 4: Shows understanding and use of equivalent fractions; makes conjectures and builds a logical progression of statements; routinely interprets mathematical results in the context of the scenario and reflects on whether the results make sense; communication is precise, using definitions clearly.
Ms. Campbell	**From Chapter 2:** Understand concepts of volume and relate volume to multiplication and to addition. Relate volume to the operations of multiplication and addition and solve real-world and mathematical problems involving volume. Recognize volume as additive. Find volumes of solid figures composed of two non-overlapping right rectangular prisms by adding the volumes of the non-overlapping parts, applying this technique to solve real-world problems.	**Level 1:** No solution is provided, or the solution has no relationship to the task; solution addresses none of the mathematical components presented in the task; inappropriate concepts are applied and/or procedures are used. **Level 2:** Solution is not complete, indicating that parts of the problem are not understood; solution addresses some, but not all, of the mathematical components presented in the task. **Level 3:** Solution shows that the student has a broad understanding of the problem and the major concepts necessary for its solution; solution addresses all of the components presented in the task. **Level 4:** Solution shows a deep understanding of the problem, including the ability to identify the appropriate mathematical concepts and the information necessary for its solution; solution completely addresses all mathematical components presented in the task; solution puts to use the underlying mathematical concepts upon which the task is designed.

Figure 5.3

intentions—to designate their students' levels of proficiency on the pathway to mastery. That said, there is no prescribed number of these levels.

Collecting Evidence of Progress Toward Mastery

To determine progress and to support the grades given to students, teachers must be able to clearly answer the question "What evidence suggests that the learners have mastered the learning or are moving toward mastery?" The evidence used to determine mastery is typically more formal than the evidence used to check for understanding. For example, an exit ticket could easily be used to determine which students mastered a given learning intention on a given day. But that may not be sufficient evidence for determining mastery of a standard or set of standards. Checks for understanding gather and provide *evidence of a learner's progress* toward a learning intention, whereas an evaluation of mastery provides *evidence that a student has demonstrated mastery* of a standard or set of standards.

The difference between checks for understanding and evaluating for mastery lies in the focus of the task, as well as the use of the evidence. In a check for understanding, teachers and students are gathering evidence about learning around specific learning intention and success criteria (see Figure 5.4).

Video 17
Scaffolding Learning in a Transfer Lesson

https://resources.corwin.com/ vlmathematics-3-5

RELATIONSHIP BETWEEN LEARNING INTENTIONS AND CHECKS FOR UNDERSTANDING

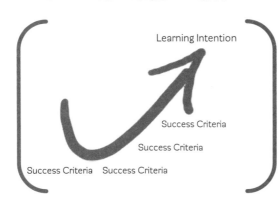

Checks for Understanding gather and provide *evidence of learners' progress* toward a learning intention using the success criteria as guides for this progression.

Figure 5.4

Ms. Buchholz, Ms. Mills, and Ms. Campbell had multiple checks for understanding throughout their lessons. In each of their classrooms, learners engaged in checks for understanding that targeted the specific learning intentions and success criteria for the lesson.

Although we can use formative assessments collected over time to evaluate mastery—evidence over time—our classrooms require single tasks that evaluate mastery (e.g., performance-based learning tasks and well-designed standardized tests). These tasks evaluate student mastery by focusing on the standard(s), asking learners to assimilate all of the learning into *a challenging mathematics task* (sometimes called a *rich mathematical task*) (Hattie et al., 2017). Again, evaluating student mastery brings together multiple concepts, procedures, and applications into a single task rather than rich tasks that target specific success criteria within a standard or standards. These tasks can include, but are not limited to, performance-based learning tasks and well-designed standardized tests (see Figure 5.5).

INGREDIENTS FOR PROGRESS TOWARD MASTERY

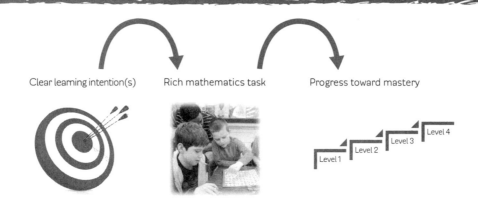

Clear learning intention(s)　　　Rich mathematics task　　　Progress toward mastery

Image source: Samarskaya/iStock.com

Figure 5.5

Figure 5.6 includes a checklist useful in creating assessment of mastery. As Ms. Showker says,

> After our team develops an assessment for mastery, we each take time to ensure that the assessment will tell us what we want and need to know about our learners. We don't want learners to walk away thinking they've got it, we think they've got it, and then find out that we were both wrong. We want evidence or proof that suggests students are mastering content or when students need more learning.

CHECKLIST FOR CREATING OR SELECTING TASKS THAT ASSESS MASTERY

All Items

❑ Is this the most appropriate type of item to use for the intended learning outcomes?

❑ Does each item or task require students to demonstrate the performance described in the specific learning outcome it measures (relevance)?

❑ Does each item present a clear and definite task to be performed (clarity)?

❑ Is each item or task presented in simple, readable language and free from excessive verbiage (conciseness)?

❑ Does each item provide an appropriate challenge (ideal difficulty)?

❑ Does each item have an answer that would be agreed upon by experts (correctness)?

❑ Is there a clear basis for awarding partial credit on items or tasks with multiple points (scoring rubric)?

❑ Is each item or task free from technical errors and irrelevant clues (technical soundness)?

❑ Is each test item free from cultural bias?

❑ Have the items been set aside for a time before reviewing them (or being reviewed by a colleague)?

Performance Items

❑ Does the item focus on learning outcomes that require complex cognitive skills and student performances?

❑ Does the task represent the content, skills, processes, and practices that are central to learning outcomes?

❑ Does the item minimize dependence on skills that are irrelevant to the intended purpose of the assessment task?

❑ Does the task provide the necessary scaffolding for students to be able to understand the task and achieve the task?

❑ Do the directions clearly describe the task?

❑ Are students aware of the basis (expectations) on which their performances will be evaluated in terms of scoring rubrics?

Source: Adapted from Linn, R. L., & Gronlund, N. E. (2000). *Measurement and assessment in teaching* (8th ed.). Upper Saddle River, NJ: Merrill Prentice Hall. Used with permission.

Figure 5.6

 This checklist is available for download at **resources.corwin.com/vlmathematics-3-5.**

> Poorly designed tasks yield poor evidence and poor decisions about where to go next.

> What separates a challenging, rich mathematics task from a rote exercise is the nature of the cognitive engagement required to complete the task.

A poorly designed task washes out the benefit of determining learner mastery. For example, a group of teachers were looking to see if learners could compare the properties of three-dimensional shapes (e.g., vertices, faces, etc.). They developed a sorting and matching task, but that did not provide them with the evidence needed to make a decision about student proficiency. We are not saying that a sorting or matching task never works; however, a sorting task at this time, for this content will not provide the evidence needed to make a decision about student proficiency. Furthermore, if the task is not engaging and relevant to our students, their level of persistence will likely skew the evidence as well. Whether the evaluation of mastery provides evidence to the teacher and student about the current level of mastery is in the nature of the task itself. In other words, poorly designed tasks yield poor evidence and poor decisions about where to go next.

In order to develop an effective evaluation that provides opportunities for learners to demonstrate mastery while at the same time provides evidence for feedback or next steps, teachers should consider the ways students can make their mathematics thinking visible. What separates a challenging, rich mathematics task from a rote exercise is the nature of the cognitive engagement required to complete the task. In mathematics exercises, learners repeat terms, concepts, ideas, procedures, or processes and apply those in novel situations.

Let us look back to the set of mastery tasks developed by Ms. Showker and her team. How learners approach these tasks and the thinking these tasks generate provides valuable information to both the teacher and the learners, allowing the learners to gain an understanding of where they are in their learning progression, identify where they need to go next in their learning, and what learning tools are needed to support this next step. What do we mean by *challenging, rich mathematics tasks*? There are many definitions:

- Accessible to all learners ("low floor, high ceiling")
- Real-life task or application
- Multiple approaches and representations
- Collaboration and discussion
- Engagement, curiosity, and creativity
- Making connections within and/or across topics and domains, vertically and horizontally

- Opportunities for extension (adapted from Boaler, 2015, 2016; Wolf, 2015)

These tasks are far different from forced-choice items that may only assess the guesswork of mathematics learners. Bringing the previous definitions to life, Antonetti and Garver (2015) reported on data from classroom walk-throughs that focused on eight features that differentiated mathematics tasks from mere rote exercises. Observers measured consistent and sustained engagement when three or more of the features were present. The eight characteristics of challenging mathematics tasks are as follows:

1. **Personal response:** Do students have the opportunity to bring their own personal experiences with mathematics to the task? Examples include any task that invites learners to bring their own background, interests, or expertise to the task. This might be an activity that provides learners with the option to create their own analogies or metaphors, allows them to select how they will share their responses to a question (e.g., writing, drawing, speaking, etc.), or lets them select the context in which a concept is explored (e.g., selection of a specific book or creation of their own problem). These examples have one thing in common: They allow learners to personalize their responses to meet their background, interests, or expertise. As we evaluate mastery, insight into how learners are making meaning of the conceptual understanding, procedural knowledge, and the application of concepts and thinking skills is important.

2. **Clear and modeled expectations:** Do learners have a clear understanding of what they are supposed to do in this mathematics task? This characteristic refers us to clear learning intentions, success criteria, learning progressions, exemplars, models, worked examples, and rubrics. (We will take an additional look at the role of rubrics later in this chapter.) Do your learners know what success looks like in this task, or are they blindly hoping to hit the end target that you have in mind for them?

3. **Sense of audience:** Do learners have a sense that this mathematics work matters to someone other than the teacher and the gradebook? Tasks that have a sense of audience mean something to individuals beyond the teacher, which provides authenticity. Sense

of audience can be established by cooperative learning or group work in which individual members have specific roles, as in a jigsaw. Other examples include community-based projects or service projects that use mathematics and contribute to the local, school, or classroom community (e.g., analyzing data from a local stream).

4. **Social interaction:** Do learners have opportunities to socially interact with their peers? Providing learners with opportunities to talk about mathematics and interact with their peers supports their meaning making and development of conceptual understanding as well as the application of concepts and thinking skills. In addition, teachers and learners get to hear other students' mathematics thinking.

5. **Emotional safety:** Do learners feel safe in asking questions or making mistakes? Even though this task seeks to evaluate the level of mastery in mathematics content, learners must still believe that they will learn from mistakes and that errors are welcomed even at this stage of their learning. To be blunt, if learners feel threatened in your mathematics classroom, they will not engage in any mathematics task.

6. **Choice:** Do learners have choices in how they access the mathematics task? As learners engage with procedures, concepts, or their application, we should offer choices of whom they work with, what materials and manipulatives are available, and what mathematics learning strategies they can use to accomplish the task. In addition, we should offer them multiple ways to show us what they know about the mathematics content.

7. **Novelty:** Does the task require learners to approach the mathematics from a unique perspective? Examples of this characteristic include engaging scenarios, discrepant events, scientific phenomena demonstrations, or games and puzzles.

8. **Authenticity:** Does the task represent an authentic learning experience, or is the experience sterile and unrealistic (e.g., a worksheet, problem-solving scenario)? We can offer learners a scenario around community and school events, have them address STEM tasks that model simple machines and require measurements, or manipulate weather data and interpret trends in weather. (adapted from Schlechty, 2011)

To evaluate the level of mastery in mathematics learning, teachers must design and implement tasks that provide opportunities for learners to truly demonstrate what they know, how they know it, and why they know it.

Ensuring Tasks Evaluate Mastery

Ms. Showker is preparing to evaluate students' mastery in understanding proportional relationships and how to graphically represent those relationships. Throughout the week, she has used checks for understanding to gather and provide evidence of her learners' progress in the following tasks:

- Identifying numerical versus categorical data
- Formulating questions to investigate
- Collecting and organizing data using surveys, polls, questionnaires, scientific experiments, and observation
- Identifying and describing line graphs and bar graphs
- Describing the relationship between the data and the best way to represent that data
- Comparing and contrasting line graphs and bar graphs
- Reading graphs and analyzing the data
- Interpreting graphs

She aligned her checks for understanding with the success criteria and specific learning intentions for each lesson. Ms. Showker's checks for understanding allow her to evaluate her students' progress and adjust their learning experiences, but they do not allow her to determine mastery of the content. Mastery assessments are used more summatively, whereas checks for understanding are used more formatively. But know that assessments are neither formative nor summative by nature; it's all in the use of the tool. And as you will see, the mastery assessments are often used to guide future learning experiences for students. Thus, they are tools that include multiple learning intentions, are typically administered at the unit level, can be used as evidence of longer-term learning, and are often used as the basis for grades. Having said that, if we really believe in mastery, grades would be updated throughout the year as students demonstrate competency of previous content. Thus, the grades for a unit taught in October might be updated when students demonstrate deeper understanding in December. For more information on competency-based or standards-based grading, see Guskey (2014). To design or select a task or possibly a cohesive set of tasks for evaluating mastery, teachers should do the following:

1. Return to the learning intentions and success criteria associated with content for which we are evaluating mastery. What is it that students were supposed to learn?

2. Create or select a challenging mathematics task (or a set of tasks) that requires learners to demonstrate their proficiency for each specific learning intention and success criterion. In other words, can students do what each of the learning intentions says they should be able to do?

3. Identify criteria for mastery and levels of progress toward mastery.

Ensuring Tests Evaluate Mastery

Tasks are great, but there will always be mathematics tests. Tests are not only common in the mathematics classroom, but they can also be an effective means for determining the mastery of learners. The intention and design of any test determines the usefulness of the evidence generated about learner mastery. Whether multiple choice or open ended, tests must provide the necessary evidence about student learning so that both the teacher and the learners can make a clear evaluation of their understanding with the specific mathematics content. In designing mathematics tests, we must take into account several aspects of that test if we are to achieve high-quality evaluation of student learning.

Whether in our own classroom or in the classrooms of the teachers featured in this book, a test designed to evaluate learner mastery must contain questions or items that are consistent with the teaching and learning in that classroom. If the focus in Ms. Mills's fourth grade classroom is on creating a number line to show comparisons and equivalencies of fractions (e.g., procedural knowledge in rational numbers), then an end-of-unit or standard test cannot contain items that solely focus on conceptual understanding or application to provide a clear evaluation of student mastery (e.g., explaining connections among representations of fractions). Likewise, if the focus in Ms. Mills's class is on the conceptual understanding of rational numbers and relationships between the multiple representations of fractions, a test for mastery cannot contain items that only focus on placing them on a number line. Therefore, the first aspect of a well-designed test is that the test items align with the expectations of the whole standard or set of standards and associated learning intentions and success criteria.

Test items should provide learners with the opportunity to demonstrate different levels of mastery. In addition to having test items that align with the expectations of the standard, a well-designed test will have questions that fall in the progression toward the standard.

In addition, the test might ask learners to explain how each approach allows them to support their solution. Including the components that build up to the standard will allow Ms. Buchholz, for example, to determine how much learners have mastered if they have not fully mastered the standard.

As we reflect on our days as elementary mathematics students, we can likely recall instances in which we missed questions on a test because we were not clear on what the questions were asking us to do. When we received feedback on the test, we may have responded to that feedback with, "Oh, that's what you wanted on number 15?" Using consistent language on a test is vital in evaluating the learning of mathematics compared to semantics. As students engage in mathematics learning, we must ensure that the language we expect them to master is the language we use in the learning experiences. For example, if Ms. Campbell plans to include questions on her test that use the term *compound shapes*, then this concept should be introduced during the learning experiences. Likewise, if she is going to use *adjacent side* or other terms for a length and width, learners need experiences with that vocabulary or terminology. Using consistent language applies to the cognitive aspects of the questions as well. We must ensure learners know what we mean by *analyze*, *explain*, or *support your answer*.

Figures 5.7 provides additional guidelines for developing well-designed tests. These checklists help to ensure that our tests provide clear evidence about our learners' mastery in mathematics.

The first aspect of a well-designed test is that the test items align with the expectations of the whole standard or set of standards and associated learning intentions and success criteria.

Teaching Takeaway

There should be items on the test that build up to the standard or mastery level.

Teaching Takeaway

Students should be familiar with the language of the test.

CHECKLISTS FOR CREATING TESTS
THAT ASSESS MASTERY

Short-Answer Items

❑ Can the items be answered with a number, symbol, word, or brief phrase?

❑ Has textbook language been avoided?

❑ Have the items been stated so that only one response is correct?

❑ Are the answer blanks equal in length (for fill-in responses)?

❑ Are the answer blanks (preferably one per item) at the end of the items, preferably after a question?

❑ Are the items free of clues (such as *a* or *an*)?

❑ Has the degree of precision been indicated for numerical answers?

❑ Have the units been indicated when numerical answers are expressed in units?

Binary (True–False) and Multiple-Binary Items

❑ Can each statement be clearly judged true or false with only one concept per statement?

❑ Have specific determiners (e.g., *usually, always*) been avoided?

❑ Have trivial statements been avoided?

❑ Have negative statements (especially double negatives) been avoided?

❑ Does a superficial analysis suggest a wrong answer?

❑ Are opinion statements attributed to some source?

❑ Are the true and false items approximately equal in length?

❑ Is there approximately an equal number of true and false items?

❑ Has a detectable pattern of answers (e.g., *T, F, T, F*) been avoided?

Matching Items

❑ Is the material for the two lists homogeneous?

❑ Is the list of responses longer or shorter than the list of premises?

❑ Are the responses brief and on the right-hand side?

❑ Have the responses been placed in alphabetical or numerical order?

❑ Do the directions indicate the basis for matching?

❑ Do the directions indicate how many times each response may be used?

❑ Are all of the matching items on the same page?

Multiple-Choice Items

☐ Does each item stem present a meaningful problem?

☐ Is there too much information in the stem?

☐ Are the item stems free of irrelevant material?

☐ Are the item stems stated in positive terms (if possible)?

☐ If used, has negative wording been given special emphasis (e.g., capitalized)?

☐ Are the distractors brief and free of unnecessary words?

☐ Are the distractors similar in length and form to the answer?

☐ Is there only one correct or clearly best answer?

☐ Are the distractors based on specific misconceptions?

☐ Are the items free of clues that point to the answer?

☐ Are the distractors and answer presented in sensible (e.g., alphabetical, numerical) order?

☐ Has *all of the above* been avoided and *none of the above* used judiciously?

☐ If a stimulus is used, is it necessary for answering the item?

☐ If a stimulus is used, does it require use of skills sought to be assessed?

Source: Adapted from Linn, R. L., & Gronlund, N. E. (2000). *Measurement and assessment in teaching* (8th ed.). Upper Saddle River, NJ: Merrill Prentice Hall. Used with permission.

Figure 5.7

 This checklist is available for download at **resources.corwin.com/vlmathematics-3-5**.

If our ultimate goal is for students to see themselves as their own mathematics teacher, we have to devote time to helping them prepare for tests. Simply telling our learners to "study" is not enough to support them in their journey to becoming assessment-capable visible learners in mathematics. As you can see, we have come full circle in this book. Ensuring that learners have clarity about the learning intentions, success criteria, and their progress toward those items will then help them prepare for

Depending on the level of proficiency demonstrated by the learner, specific, constructive, and timely feedback supports learners as they—together with the teacher—evaluate where they are going, how they are going, and where they are going next

Task feedback addresses how well the task has been performed—correct or incorrect.

this evaluation of mastery. Providing learners with opportunities to connect the learning intentions and success criteria to the type of question they will likely see on a test encourages them to take ownership of their mathematics learning.

Feedback for Mastery

With the learning intention clear, a definition of *success* established, and a challenging mathematics assessment of mastery developed and implemented, the next key item is feedback. The nature of the feedback on learners' performance is an essential and necessary component in the Visible Learning mathematics classroom. Depending on the level of proficiency demonstrated by the learner, specific, constructive, and timely feedback supports learners as they—together with the teacher—evaluate where they are going, how they are going, and where they are going next (see Hattie & Timperley, 2007).

Task Feedback

For learners at the earliest level of mastery, **task feedback** develops student understanding of specific procedures, concepts, and applications. This type of feedback is corrective, precise, and focused on the accuracy of the learners' responses to the mastery task. For example, Ms. Showker may provide written or verbal feedback that says, "Take a look at your axes. The horizontal axis should contain your categories for that type of bar graph." She may indicate to a learner that a specific question is wrong and needs revisiting before moving on in the learning. On the other hand, she may point out, "You have correctly labeled your axes based on the data you collected. Now move on to graphing your data."

Learners rely on task feedback to add additional structure to their conceptual understanding, procedural knowledge, and application of concepts and thinking skills. This may include examples and non-examples, additional learning on procedural steps, and contexts of the task. Ms. Showker may sit down with a learner who has missed a specific question and provide additional examples for deciding whether data are numerical or categorical and then choosing a scale. She may even provide two scenarios and ask the learner to compare and contrast them to clarify understanding. Each learner's successful assimilation of feedback,

ELEMENTS OF EFFECTIVE FEEDBACK

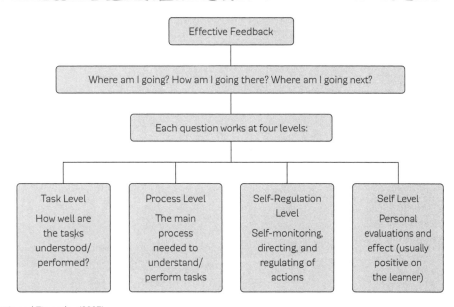

Source: Hattie and Timperley (2007).

Figure 5.8

and thus use of the feedback to decide where to go next, rests solely on whether each learner understands what the feedback means and how he or she can use it to move forward with mathematics learning. Effective feedback (Figure 5.8) and effective use of that feedback supports this initial learning.

Process Feedback

As learners begin to develop proficiency with specific content, ideas, and terms, the feedback should increasingly shift to process feedback. **Process feedback** is critical as learners explore the *why* and the *how* of specific mathematics content. In their initial assessment

Teaching Takeaway

Feedback should answer three questions for the learner: Where am I going? How am I going? Where am I going next?

Process feedback focuses on the strategies needed to perform the task.

of mastery, learners received and assimilated task feedback into their work to develop a deeper understanding of procedures, concepts, and applications. To move learners beyond what is simply right or wrong, example or non-example, they must receive and incorporate feedback that focuses on the processes or strategies associated with accomplishing the specific task. Returning to Ms. Showker's classroom, she may not indicate whether a particular response is correct or incorrect, but she might simply ask, "Why do you believe that this is the best way to represent your data? What information can you infer from the graph? Do you have any data that would allow you to verify these predictions?"

Whether from the teacher or peers, learners should receive feedback on their thinking, not just the accuracy of their response. For example, teachers might engage students in further dialogue about the use of specific strategies to solve a particular problem. Again, this feedback can come from the teacher or their peers. For example, Figure 5.9 shows an example of peer-assisted reflection (PAR) (Almarode et al., 2019). In this scenario, learners complete their task—along with their mathematics story that explains the process for completing the task (not just *what* they did, by *why* they did it)—that is ready to be reviewed by a peer. The peer feedback is offered in two phases. First, peers provide each other written feedback in the form of annotations and a rating toward mastery of each success criterion during a silent review phase. Second, peers discuss the written feedback they provided and ask any clarifying questions they might have about that feedback. The final step for students is to revise their draft solution into a final submission and include a reflection of how their thinking changed throughout this process.

Teaching Takeaway

To provide the most amount of feedback to the greatest number of learners as possible, incorporate student-to-student feedback and strategies for student-to-self feedback.

The PAR cycle gives students the opportunity to compare and contrast: *This is what I used to be able to do; this is what I can do now. This is how I used to think about this problem; this is how I think about it now. This is what I used to know; this is what I know now.* In addition to these before-and-after snapshots, the feedback and annotation components of PARs can collect much of the connective tissue that bridged students from where they were to where they are. In other words, not only does growth as an outcome become blatant to students, but students become aware of their own growth process as well.

Process feedback supports making connections, use of multiple strategies, self-explanation, self-monitoring, self-questioning, and critical thinking. For example, Ms. Showker may ask the learner what strategies he or she used in making the decisions about increasing or decreasing intervals and ask if the strategy worked well or if a different strategy may be more efficient. Rather than focusing solely on the correct answer regarding the relationship between an independent and a dependent variable, a teacher may ask a student, "What is your explanation for your answer?" The focus of process feedback is on the relationships between ideas, students' strategies for evaluating the reasonableness of an answer or solution, explicitly learning from mistakes, and helping the learner identify different strategies for addressing a task.

Like task feedback, process feedback should be specific and constructive and should support learners' pathways toward self-regulation feedback. That is, the feedback should deepen thinking, reasoning, explanations, and connections. Does the teacher prompt learners through strategic questioning related to the learning process? What appears to be wrong, and why? What approach or strategies did the learner use or apply to the task? What is an explanation for the answer, response, or solution? What are the relationships with other parts of the task?

Self-Regulation Feedback

Self-regulation feedback is the learner knowing what to do when she or he approaches a new and different problem, is stuck, or has to apply concepts and thinking in a new way. Learners who have reached a deep level of conceptual understanding and are armed with multiple strategies are equipped to self-regulate as they transfer their learning to more rigorous tasks. Highly proficient learners benefit from self-regulation feedback, although this is not the only type of feedback that is important to these learners. For example, when teachers detect a misconception or notice a gap in foundational or background learning, learners benefit from both task and process feedback in these situations. However, a majority of the feedback at this part of the learning process should be self-regulation through metacognition. The teacher's role in the feedback at this level is to ask questions to prompt further metacognition.

> EFFECT SIZE FOR ASSESSMENT-CAPABLE VISIBLE LEARNERS = **1.33**

> **Self-regulation feedback** involves the learner self-monitoring his or her own progress toward a specific goal.

> EFFECT SIZE FOR SELF-VERBALIZATION AND SELF-QUESTIONING = **0.55**

PEER-ASSISTED REFLECTION FOR GRAPHING TASK

Success Criteria

[ ] I can formulate questions necessary for completing the task.

[ ] I can collect and organize the data generated from those questions.

[ ] I can represent the data in a line graph or bar graph.

[ ] I can justify my choice of graph for representing the data.

[ ] I can interpret my graph to address the specific task.

[ ] I can write an explanation of the process in my mathematics notebook.

Scenarios (circle the option selected by your peer)

1. Memorial Elementary is changing the mascot of our school. Your task is to survey students throughout the school so that you can make a recommendation about the type of mascot and his or her specific characteristics (e.g., type of animal, color, and name).

2. The local fast-food restaurant is trying to offer healthier side options in the "kid's meals." Your task is to poll students in the school so that you can make a recommendation about side options that are both appealing to your peers and healthy (e.g., types of fruits, types of veggies, mixture of the two, etc.).

3. The principal is creating a safety manual for the school that will help prepare teachers and students for different types of severe weather or natural disasters. Your task is to provide data that help her decide how to prioritize those events in the safety manual (e.g., which one should be listed first, which one is not likely to occur, etc.).

Reviewed by: _____

Rate your peer's mastery of the success criterion (this is the *last* thing you do):

[ ] I can formulate questions necessary for completing the task.

0—DO NOT check that box	1—ALMOST check that box	2—CHECK that box
None of the questions will generate the data necessary to address the task.	Some of the questions will generate the data necessary to address the task.	All of the questions will generate the data necessary to address the task.

[ ] I can collect and organize the data generated from those questions.

0—DO NOT check that box	1—ALMOST check that box	2—CHECK that box

[ ] I can represent the data in a line graph or bar graph.

0—DO NOT check that box	1—ALMOST check that box	2—CHECK that box

[ ] I can justify my choice of graph for representing the data.

0—DO NOT check that box	1—ALMOST check that box	2—CHECK that box

[ ] I can interpret my graph to address the specific task.

0—DO NOT check that box	1—ALMOST check that box	2—CHECK that box

[ ] I can write an explanation of the process in my mathematics notebook.

0—DO NOT check that box	1—ALMOST check that box	2—CHECK that box

DRAFT SOLUTION

ANNOTATIONS (author's and peer's)

REVISED SOLUTION

ANNOTATIONS (author only)

Figure 5.9

EFFECT SIZE FOR
EVALUATION AND
REFLECTION = 0.75

The ability to think about your own thinking promotes learners' self-awareness, problem solving around the learning task, and understanding of what they need to do to complete the task.

Eventually, learners practice metacognition independently through self-verbalization, self-questioning, and self-reflection. Ms. Showker recalls a student working diligently on the first part of the mastery task at the beginning of this chapter. Midway through the task, the learner stopped and began to erase his work, stating, "This answer does not make sense with the picture, and I can't think of a scenario where my answer works. It must be wrong." Learners take personal ownership of their learning, which provides increased motivation and understanding. This has been and continues to be a well-documented finding in education research (e.g., National Research Council, 2000). The ability to think about your own thinking promotes learners' self-awareness, problem solving around the learning task, and understanding of what they need to do to complete the task.

To reiterate, assessment-capable visible learners know what to do when they get stuck, when a new challenge arises, and when their teacher may not be available to help. This is self-regulation feedback.

Conclusion

Over the course of this book, we set out to portray the teaching of mathematics in the Visible Learning classroom. This brought together three elements of mathematics learning (conceptual understanding, procedural knowledge, and the application of concepts and thinking skills) with three phases of learning: surface, deep, and transfer.

Visible mathematics learning is an attainable goal when mathematics teachers *see* learning through the eyes of their students and students *see* themselves as their own mathematics teachers. Together, this type of learning environment develops assessment-capable visible learners. These learners can do the following:

1. Know their current level of mathematics learning

2. Know where they are going next in meeting their current mathematics learning goals and are confident to take on the challenge

3. Select the most appropriate tools, problem-solving approaches, and skills to guide their learning

4. Seek feedback and recognize errors are opportunities to enhance their mathematics learning

5. Monitor their progress and adjust their mathematics learning

6. Recognize their learning and support their peers in their own mathematics learning journey

Teaching mathematics in the Visible Learning classroom demands as much from the teacher as from the learner. We have to create a learning environment that promotes clarity in learning, provides challenging mathematics tasks, checks for understanding, and enables a clear evaluation of mastery. We must know our impact on learning! Yes, there will be days that are better than others. Learning will be stronger with some content than other content. On the most successful days, celebrate the learning that your students do. On days when there is a less-than-desirable impact on student learning, stay focused on the main thing. We keep the main thing *the main thing* by recalibrating our mindframes about teaching and learning in the mathematics classroom. We can do this by asking ourselves these recalibrating questions:

1. What do I want my students to learn?

2. What evidence will convince me that they have learned it?

3. How will I check learners' understanding and progress?

4. What tasks will get my students to mastery?

5. How will I adjust the rigor of the tasks to meet the needs of all learners?

6. What resources do I need?

7. How will I manage the learning?

The classrooms of Ms. Buchholz, Ms. Mills, Ms. Campbell, and Ms. Showker do just that daily (Figures 5.10 through 5.12).

MS. BUCHHOLZ'S VISIBLE LEARNING IN THE MATHEMATICS CLASSROOM

Conceptual Understanding: the meaning of multiplication as a one-to-many constant relationship between two sets.

Procedural Knowledge: evaluation of fluent and nonfluent division strategies based on the representation and the numbers.

Application: strategies for multiplication and dividing small numbers can be revised to multiply and divide large numbers.

Figure 5.10

MS. MILLS'S VISIBLE LEARNING IN THE MATHEMATICS CLASSROOM

Conceptual Understanding: the connections among representations of fractions.

Procedural Knowledge: strategies for comparing fractions that rely on the value of fractions as numbers.

Application: the relationship between equivalent fractions and decimals.

Figure 5.11

MS. CAMPBELL'S VISIBLE LEARNING IN THE MATHEMATICS CLASSROOM

Conceptual Understanding: what is volume—volume is the amount of space inside a solid figure.

Procedural Knowledge: computing volume of right rectangular prisms based on edge lengths and the area of one face.

Application: solving problems involving the packing of three dimensional shapes.

Figure 5.12

Final Reflection

Summarizing the content in this book, reflect on the following questions:

1. Using a specific standard or standards for an upcoming unit, describe what mastery would look like for that content.

2. How will you check for understanding as your learners progress toward mastery? How will these checks be different from your evaluation of their mastery of the standard or standards?

3. How do you plan to evaluate mastery of this particular content—task, test, or both?

4. Reflect on a recent mathematics task in your classroom. Using the definition and characteristics from this chapter, does it "qualify" as a challenging mathematics task?

5. Explain the role of feedback in supporting your learners' journey to mastery.

Appendix A

Effect Sizes

The Visible Learning research synthesizes findings from **1,800** meta-analyses of **80,000** studies involving **300** million students, into what works best in education.

STUDENT		ES
Prior knowledge and background		
Field independence	◐	0.68
Non-standard dialect use	○	−0.29
Piagetian programs	●	1.28
Prior ability	●	0.94
Prior achievement	◐	0.55
Relating creativity to achievement	◐	0.40
Relations of high school to university achievement	◐	0.60
Relations of high school achievement to career performance	●	0.38
Assessment-capable visible learners	●	1.33
Working memory strength	◐	0.57
Beliefs, attitudes, and dispositions		
Attitude to content domains	●	0.35
Concentration/persistence/engagement	◐	0.56
Grit/incremental vs. entity thinking	●	0.25
Mindfulness	●	0.29
Morning vs. evening		0.12
Perceived task value	◐	0.46
Positive ethnic self-identity		0.12
Positive self-concept	◐	0.41
Self-efficacy	●	0.92
Stereotype threat	○	−0.33
Student personality attributes	●	0.26
Motivational approach, orientation		
Achieving motivation and approach	◐	0.44
Boredom	○	−0.49
Deep motivation and approach	◐	0.69
Depression	○	−0.36
Lack of stress		0.17
Mastery goals		0.06
Motivation	◐	0.42
Performance goals	○	−0.01
Reducing anxiety	◐	0.42
Surface motivation and approach	○	−0.11
Physical influences		
ADHD	○	−0.90
ADHD – treatment with drugs	●	0.32
Breastfeeding		0.04
Deafness	○	−0.61
Exercise/relaxation	●	0.26
Gender on achievement		0.08
Lack of illness	●	0.26
Lack of sleep	○	−0.05
Full compared to pre-term/low birth weight	◐	0.57
Relative age within a class	◐	0.45

CURRICULA		ES
Reading, writing, and the arts		
Comprehensive instructional programs for teachers	●	0.72
Comprehension programs	◐	0.47
Drama/arts programs	●	0.38
Exposure to reading	◐	0.43
Music programs	●	0.37
Phonics instruction	●	0.70
Repeated reading programs	●	0.75
Second/third chance programs	◐	0.53
Sentence combining programs		0.15
Spelling programs	◐	0.58
Visual-perception programs	◐	0.55
Vocabulary programs	◐	0.62
Whole language approach		0.06
Writing programs	◐	0.45
Math and sciences		
Manipulative materials on math	●	0.30
Mathematics programs	◐	0.59
Science programs	◐	0.48
Use of calculators	●	0.27
Other curricula programs		
Bilingual programs	●	0.36
Career interventions	●	0.38
Chess instruction	●	0.34
Conceptual change programs	●	0.99
Creativity programs	◐	0.62
Diversity courses		0.09
Extra-curricula programs	●	0.20
Integrated curricula programs	◐	0.47
Juvenile delinquent programs		0.12
Motivation/character programs	●	0.34
Outdoor/adventure programs	◐	0.43
Perceptual-motor programs		0.08
Play programs	◐	0.50
Social skills programs	●	0.39
Tactile stimulation programs	◐	0.58

Access the complete and most recent versions of the influence chart at: https://www.visiblelearningplus.com/content/research-john-hattie

HOME		ES
Family structure		
Adopted vs. non-adopted care	●	0.25
Engaged vs. disengaged fathers	●	0.20
Intact (two-parent) families	●	0.23
Other family structure	○	0.16
Home environment		
Corporal punishment in the home	○	−0.33
Early years' interventions	◐	0.44
Home visiting	●	0.29
Moving between schools	○	−0.34
Parental autonomy support	○	0.15
Parental involvement	◐	0.50
Parental military deployment	○	−0.16
Positive family/home dynamics	◐	0.52
Television	○	−0.18
Family resources		
Family on welfare/state aid	○	−0.12
Non-immigrant background	○	0.01
Parental employment	○	0.03
Socio-economic status	◐	0.52

SCHOOL		ES
Leadership		
Collective teacher efficacy	●	1.57
Principals/school leaders	●	0.32
School climate	●	0.32
School resourcing		
External accountability systems	●	0.31
Finances	●	0.21
Types of school		
Charter schools	○	0.09
Religious schools	●	0.24
Single-sex schools	○	0.08
Summer school	●	0.23
Summer vacation effect	○	−0.02
School compositional effects		
College halls of residence	○	0.05
Desegregation	●	0.28
Diverse student body	○	0.10
Middle schools' interventions	○	0.08
Out-of-school curricula experiences	●	0.26
School choice programs	○	0.12
School size (600–900 students at secondary)	◐	0.43
Other school factors		
Counseling effects	●	0.35
Generalized school effects	◐	0.48
Modifying school calendars/timetables	○	0.09
Pre-school programs	●	−0.28
Suspension/expelling students	○	−0.20

CLASSROOM		ES
Classroom composition effects		
Detracking	○	0.09
Mainstreaming/inclusion	●	0.27
Multi-grade/age classes	○	0.04
Open vs. traditional classrooms	○	0.01
Reducing class size	●	0.21
Retention (holding students back)	○	−0.32
Small group learning	◐	0.47
Tracking/streaming	○	0.12
Within class grouping	○	0.18
School curricula for gifted students		
Ability grouping for gifted students	●	0.30
Acceleration programs	◐	0.68
Enrichment programs	◐	0.53
Classroom influences		
Background music	○	0.10
Behavioral intervention programs	◐	0.62
Classroom management	●	0.35
Cognitive behavioral programs	●	0.29
Decreasing disruptive behavior	●	0.34
Mentoring	○	0.12
Positive peer influences	◐	0.53
Strong classroom cohesion	◐	0.44
Students feeling disliked	○	−0.19

TEACHER		ES
Teacher attributes		
Average teacher effects	●	0.32
Teacher clarity	●	0.75
Teacher credibility	●	0.90
Teacher estimates of achievement	●	1.29
Teacher expectations	◐	0.43
Teacher personality attributes	◐	0.23
Teacher performance pay	○	0.05
Teacher verbal ability	●	0.22
Teacher-student interactions		
Student rating of quality of teaching	◐	0.50
Teachers not labeling students	◐	0.61
Teacher-student relationships	◐	0.52
Teacher education		
Initial teacher training programs	○	0.12
Micro-teaching/video review of lessons	●	0.88
Professional development programs	◐	0.41
Teacher subject matter knowledge	○	0.11

Key for rating

● Potential to considerably accelerate student achievement

◐ Potential to accelerate student achievement

● Likely to have positive impact on student achievement

◌ Likely to have small positive impact on student achievement

○ Likely to have a negative impact on student achievement

ES Effect size calculated using Cohen's *d*

corwin.com/visiblelearning

Access the complete and most recent versions of the influence chart at: https://www.visiblelearningplus.com/content/research-john-hattie

The Visible Learning research synthesizes findings from **1,800** meta-analyses of **80,000** studies involving **300** million students, into what works best in education.

TEACHING: Focus on student learning strategies

		ES
Strategies emphasizing student meta-cognitive/self-regulated learning		
Elaboration and organization	●	0.75
Elaborative interrogation	○	0.42
Evaluation and reflection	●	0.75
Meta-cognitive strategies	○	0.60
Help seeking	●	0.72
Self-regulation strategies	○	0.52
Self-verbalization and self-questioning	○	0.55
Strategy monitoring	○	0.58
Transfer strategies	●	0.86
Student-focused interventions		
Aptitude/treatment interactions		0.19
Individualized instruction	●	0.23
Matching style of learning	●	0.31
Student-centered teaching	●	0.36
Student control over learning	○	0.02
Strategies emphasizing student perspectives in learning		
Peer tutoring	○	0.53
Volunteer tutors	●	0.26
Learning strategies		
Deliberate practice	●	0.79
Effort	●	0.77
Imagery	○	0.45
Interleaved practice	●	0.21
Mnemonics	●	0.76
Note taking	○	0.50
Outlining and transforming	○	0.66
Practice testing	○	0.54
Record keeping	○	0.52
Rehearsal and memorization	●	0.73
Spaced vs. mass practice	○	0.60
Strategy to integrate with prior knowledge	●	0.93
Study skills	○	0.46
Summarization	●	0.79
Teaching test taking and coaching	●	0.30
Time on task	○	0.49
Underlining and highlighting	○	0.50

TEACHING: Focus on teaching/instructional strategies

		ES
Strategies emphasizing learning intentions		
Appropriately challenging goals	○	0.59
Behavioral organizers	○	0.42
Clear goal intentions	○	0.48
Cognitive task analysis	●	1.29
Concept mapping	○	0.64
Goal commitment	○	0.40
Learning goals vs. no goals	○	0.68
Learning hierarchies-based approach		0.19
Planning and prediction	●	0.76
Setting standards for self-judgment	○	0.62
Strategies emphasizing success criteria		
Mastery learning	○	0.57
Worked examples	●	0.37
Strategies emphasizing feedback		
Classroom discussion	●	0.82
Different types of testing		0.12
Feedback	●	0.70
Providing formative evaluation	○	0.48
Questioning	○	0.48
Response to intervention	●	1.29
Teaching/instructional strategies		
Adjunct aids	●	0.32
Collaborative learning	●	0.34
Competitive vs. individualistic learning	●	0.24
Cooperative learning	○	0.40
Cooperative vs. competitive learning	○	0.53
Cooperative vs. individualistic learning	○	0.55
Direct/deliberate instruction	○	0.60
Discovery-based teaching	●	0.21
Explicit teaching strategies	○	0.57
Humor		0.04
Inductive teaching	○	0.44
Inquiry-based teaching	○	0.40
Jigsaw method	●	1.20
Philosophy in schools	○	0.43
Problem-based learning	●	0.26
Problem-solving teaching	○	0.68
Reciprocal teaching	●	0.74
Scaffolding	●	0.82
Teaching communication skills and strategies	○	0.43

Access the complete and most recent versions of the influence chart at: https://www.visiblelearningplus.com/content/research-john-hattie

TEACHING: Focus on implementation method		ES
Implementations using technologies		
Clickers	●	0.22
Gaming/simulations	●	0.35
Information communications technology (ICT)	○	0.47
Intelligent tutoring systems	○	0.48
Interactive video methods	○	0.54
Mobile phones	●	0.37
One-on-one laptops		0.16
Online and digital tools	●	0.29
Programmed instruction	●	0.23
Technology in distance education		0.01
Technology in mathematics	●	0.33
Technology in other subjects	○	0.55
Technology in reading/literacy	●	0.29
Technology in science	●	0.23
Technology in small groups	●	0.21
Technology in writing	○	0.42
Technology with college students	○	0.42
Technology with elementary students	○	0.44
Technology with high school students	●	0.30
Technology with learning needs students	○	0.57
Use of PowerPoint	●	0.26
Visual/audio-visual methods	●	0.22
Web-based learning		0.18
Implementations using out-of-school learning		
After-school programs	○	0.40
Distance education		0.13
Home-school programs		0.16
Homework	●	0.29
Service learning	○	0.58
Implementations that emphasize school-wide teaching strategies		
Co- or team teaching		0.19
Interventions for students with learning needs	●	0.77
Student support programs – college	●	0.21
Teaching creative thinking	●	0.34
Whole-school improvement programs	●	0.28

Key for rating

- ◉ Potential to considerably accelerate student achievement
- ◯ Potential to accelerate student achievement
- ● Likely to have positive impact on student achievement
- Likely to have small positive impact on student achievement
- ○ Likely to have a negative impact on student achievement
- ES Effect size calculated using Cohen's *d*

Appendix B

Planning for Clarity Guide

Teaching for Clarity PLANNING GUIDE

ESTABLISHING PURPOSE

1   What are the key content standards I will focus on in this lesson?

2   What are the learning intentions (the goal and *why* of learning, stated in student-friendly language) I will focus on in this lesson?

Content:

Language:

Social:

3 When will I introduce and reinforce the learning intention(s) so that students understand it, see the relevance, connect it to previous learning, and can clearly communicate it themselves?

SUCCESS CRITERIA

4 What evidence shows that students have mastered the learning intention(s)? What criteria will I use?

I can statements:

5 How will I check students' understanding (assess learning) during instruction and make accommodations?

INSTRUCTION

6 What activities and tasks will move students forward in their learning?

7 What resources (materials and sentence frames) are needed?

8   How will I organize and facilitate the learning? What questions will I ask? How will I initiate closure?

Appendix C

Learning Intentions and Success Criteria Template

Learning Intentions	Conceptual Understanding	Procedural Knowledge	Application of Concepts and Thinking Skills
Unistructural (one idea)			
Multistructural (many ideas)			
Relational (related ideas)			
Extended abstract (extending ideas)			

Success Criteria	Conceptual Understanding	Procedural Knowledge	Application of Concepts and Thinking Skills
Unistructural (one idea)			
Multistructural (many ideas)			
Relational (related ideas)			
Extended abstract (extending ideas)			

 This template is available for download at **resources.corwin.com/vlmathematics-3-5**.

A Selection of International Mathematical Practice or Process Standards*

*Note that this is a non-exhaustive list of international mathematical practice/process standards as of June 2016. Because standards are often under review, you can look to your own state or country's individual documents to find the most up-to-date practice/process standards.

USA Common Core State Standards 8 Mathematical Practices[a]	USA Texas Essential Knowledge and Skills TEKS 7 Mathematical Practice Standards[b]	USA Virginia Mathematics 5 Standards of Learning[c]	International Baccalaureate 6 Assessment Objectives[d]	Hong Kong Key Learning Area 7 Generic Skills[e]	Singapore Mathematical Problem-Solving Processes[f]	Australian F-10 Mathematics Curriculum Key Ideas[g]
1. Make sense of problems and persevere in solving them.	A. Apply mathematics to problems arising in everyday life, society, and the workplace.	Mathematical problem solving	Knowledge and understanding	Collaboration skills	Reasoning, communications, and connections	Understanding
2. Reason abstractly and quantitatively.	B. Use a problem-solving model that incorporates analyzing given information, formulating a plan or strategy, determining a solution, justifying the solution, and evaluating the problem-solving process and the reasonableness of the solution.	Mathematical communication	Problem solving	Communication skills	Applications and modeling	Fluency
3. Construct viable arguments and critique the reasoning of others.		Mathematical reasoning	Communication and interpretation	Creativity	Thinking skills and heuristics	Problem solving
4. Use appropriate tools strategically.	C. Select tools, including real objects, manipulatives, paper and pencil, and technology as appropriate.	Mathematical connection	Technology	Critical-thinking skills		Reasoning
5. Attend to precision.	D. Communicate mathematical ideas, reasoning, and their implications using multiple representations, including symbols, diagrams, graphs, and language as appropriate.	Mathematical representations	Reasoning	Information technology skills		
6. Look for and make use of structure.			Inquiry approaches	Numeracy skills		
				Problem-solving skills		

(Continued)

(Continued)

USA Common Core State Standards 8 Mathematical Practices[a]	USA Texas Essential Knowledge and Skills TEKS 7 Mathematical Practice Standards[b]	USA Virginia Mathematics 5 Standards of Learning[c]	International Baccalaureate 6 Assessment Objectives[d]	Hong Kong Key Learning Area 7 Generic Skills[e]	Singapore Mathematical Problem-Solving Processes[f]	Australian F-10 Mathematics Curriculum Key Ideas[g]
7. Look for and express regularity in repeated reasoning.	E. Create and use representations to organize, record, and communicate mathematical ideas. F. Analyze mathematical relationships to connect and communicate mathematical ideas.					
8. Model with mathematics.	G. Display, explain, and justify mathematical ideas and arguments using precise mathematical language in written or oral communication.					

[a]Retrieved June 22, 2016, from http://www.corestandards.org/Math/Practice/.

[b]Retrieved June 22, 2016, from http://ritter.tea.state.tx.us/rules/tac/chapter111/ch111a.html.

[c]Retrieved June 22, 2016, from http://www.doe.virginia.gov/testing/sol/standards_docs/mathematics/2009/stds_math.pdf.

[d]Retrieved June 22, 2016, from http://www.ibo.org/globalassets/publications/recognition/5_maths.pdf.

[e]Retrieved June 22, 2016, from http://www.edb.gov.hk/attachment/en/curriculum-development/kla/ma/curr/Math_CAGuide_e_2015.pdf.

[f]Retrieved June 22, 2016, from https://www.moe.gov.sg/docs/default-source/document/education/syllabuses/sciences/files/mathematics-syllabus-(primary-1-to-4).pdf.

[g]Retrieved June 22, 2016, from http://www.australiancurriculum.edu.au/mathematics/curriculum/f-10?layout=1.

Source: Standards for Mathematical Practice, CCSSO.

References

Almarode, J. T., Fisher, D., Assof, J., Hattie, J. A., & Frey, N. (2019). *Teaching mathematics in the Visible Learning classroom, Grades 9–12*. Thousand Oaks, CA: Corwin.

American Psychological Association, Coalition for Psychology in Schools and Education. (2015). *Top 20 principles from psychology for preK–12 teaching and learning*. Retrieved from http://www.apa.org/ed/schools/cpse/top-twenty-principles.pdf

Anno, M. (1983). *Anno's mysterious multiplying jar*. New York, NY: Philomel Books.

Antonetti, J., & Garver, J. (2015). *17,000 classroom visits can't be wrong*. Alexandria, VA: Association for Supervision and Curriculum Development.

Berry, R. Q., III, & Thunder, K. (2017). Concrete, representational, and abstract: Building fluency from conceptual understanding. *Virginia Mathematics Teacher, 43*(2), 28–32.

Biggs, J. B., & Collis, K. F. (1982). *Evaluating the quality of learning: The SOLO taxonomy (structure of observed learning outcome)*. New York, NY: Academic Press.

Boaler, J. (2015). *What's math got to do with it? How teachers and parents can transform mathematics learning and inspire success* (Rev. ed.). New York, NY: Penguin.

Boaler, J. (2016). *Mathematical mindsets*. New York, NY: Jossey-Bass.

Clements, D. H., & Sarama, J. (2007). Early childhood mathematics learning. In F. K. Lester, Jr. (Ed.), *Second handbook of research on mathematics teaching and learning* (pp. 461–555). New York, NY: Information Age.

Curriculum Services Canada. (2011). *Bansho (board writing)* (Special Ed. 17). Ontario, Canada: Author.

Empson, S. B., & Levi, L. (2011). *Extending children's mathematics: Fractions and decimals: Innovations in cognitively guided instruction*. Portsmouth, NH: Heinemann.

Fennell, F. S., Kobett, B. M., & Wray, J. A. (2017). *The formative 5: Everyday assessment techniques for every math classroom*. Thousand Oaks, CA: Corwin.

Fisher, D., & Frey, N. (2008). Homework and the gradual release of responsibility: Making "responsibility" possible. *English Journal, 98*(2), 40–45.

Frey, N., Hattie, J., & Fisher, D. (2018). *Developing assessment-capable visible learners*. Thousand Oaks, CA: Corwin.

Gagnon, J. C., & Maccini, P. (2001). Preparing students with disabilities for algebra. *Teaching Exceptional Children, 34*(1), 8–15.

Guskey, T. R. (2014). *On your mark: Challenging the conventions of grading and reporting.* Bloomington, IN: Solution Tree.

Hattie, J. (2009). *Visible learning: A synthesis of over 800 meta-analyses relating to achievement.* New York, NY: Routledge.

Hattie, J., Fisher, D., Frey, N., Gojak, L. M., Moore, S. D., & Mellman, W. (2017). *Visible learning for mathematics: What works best to optimize student learning.* Thousand Oaks, CA: Corwin.

Hattie, J., & Timperley, H. (2007). The power of feedback. *Review of Educational Research, 77*(1), 81–112.

Hattie, J., & Zierer, K. (2018). *10 mindframes for visible learning: Teaching for success.* New York, NY: Routledge.

Hook, P., & Mills, J. (2011). *SOLO taxonomy: A guide for schools. Book 1.* Laughton, United Kingdom: Essential Resources.

Humphreys, C., & Parker, R. (2015). *Making number talks matter: Developing mathematical practices and deepening understanding, Grades 4–10.* Portland, ME: Stenhouse Publishers.

Jacobs, V. R., Lamb, L. L. C., & Philipp, R. A. (2010). Professional noticing of children's mathematical thinking. *Journal for Research in Mathematics Education, 41*(2), 169–202.

Jansen, A., Cooper, B., Vascellaro, S., & Wandless, P. (2016). Rough-draft talk in mathematics classrooms. *Mathematics Teaching in the Middle School, 22*(5), 304–307.

Kuehnert, E. R. A., Eddy, C. M., Miller, D., Pratt, S. S., & Senawongsa, C. (2018). Bansho: Visually sequencing mathematical ideas. *Teaching Children Mathematics, 24*(6), 362–369.

Linn, R. L., & Gronlund, N. E. (2000). *Measurement and assessment in teaching* (8th ed.). Upper Saddle River, NJ: Merrill Prentice Hall.

National Council of Teachers of Mathematics. (1991). Professional standards for teaching mathematics. Retrieved from http://www.nctm.org/standards/content.aspx?id=26628

National Council of Teachers of Mathematics. (2014). *Principles to actions: Ensuring mathematical success for all.* Reston, VA: Author.

National Governors Association Center for Best Practices, Council of Chief State School Officers. (2010). *Common Core State Standards for Mathematics.* Washington, DC: Author.

National Reading Panel. (2000). *Teaching children to read: An evidence-based assessment of the scientific research literature on reading and its implications for reading instruction.* Donald N. Langenburg (Ed.). National Institute of Child Health and Human Development. Washington, DC: U.S. Government Printing Office.

National Research Council. (2000). *How people learn: Brain, mind, experience, and school* (Expanded ed.) Washington, DC: National Academies Press.

National Research Council. (2009). *Mathematics learning in early childhood: Paths toward excellence and equity*. Committee on Early Childhood Mathematics, Christopher T. Cross, Taniesha A. Woods, and Heidi Schweingruber (Eds.). Center for Education, Division of Behavioral and Social Sciences and Education. Washington, DC: National Academy Press.

Parrish, S. (2014). *Number talks: Whole number computation*. Sausalito, CA: Math Solutions.

Schlechty, P. C. (2011). *Working on the work. An action plan for teachers, principals, and superintendents*. San Francisco, CA: Jossey-Bass.

Small, M. (2012). *Good questions: Great ways to differentiate mathematics instruction* (2nd ed.) Reston, VA: National Council of Teachers of Mathematics.

Smith, M. S., & Stein, M. K. (2011). *5 practices for orchestrating productive mathematics discussions*. Reston, VA: National Council of Teachers of Mathematics.

Suntex International Inc. (1988). *24 Game*. Easton, PA: Author.

Thunder, K. (2014). *Differentiating elementary mathematics instruction*. Presentation, Harrisonburg, VA.

Thunder, K., & Demchak, A. N. (2012). *Using literacy strategies to gain deep mathematical understanding in Grades 2–5*. Presentation, Roanoke, VA.

Thunder, K., & Demchak, A. N. (2016). The math diet: An instructional framework to grow mathematicians. *Teaching Children Mathematics, 22*(7), 389–392.

Thunder, K., & Demchak, A. N. (2017). *Workshop models for literacy and math, Grades 1–4*. Presentation for Virginia School/University Partnership, Charlottesville, VA.

Trocki, A., Taylor, C., Starling, T., Sztajn, P., & Heck, D. (2014/2015). Launching a discourse-rich mathematics lesson. *Teaching Children Mathematics, 21*(5), 276–281.

Van de Walle, J. A., Karp, K. S., & Bay-Williams, J. M. (2018). *Elementary and middle school mathematics: Teaching developmentally* (10th ed.). New York, NY: Pearson Education.

Virginia Department of Education. (2016). *Mathematics 2016 standards of learning*. Richmond, VA: Author.

Wolf, N. B. (2015). *Modeling with mathematics. Authentic problem solving in middle school*. Portsmouth, NH: Heinemann Publishing.

Index

All students should have the opportunity to be successful!

Visible Learning^{plus} is based on one simple belief: Every student should experience at least one year's growth over the course of one school year. Visible Learning^{plus} translates the groundbreaking Visible Learning research by professor John Hattie into a practical model of inquiry and evaluation. Bring Visible Learning to your daily classroom practice with these additional resources across mathematics, literacy, and science.

John Hattie, Douglas Fisher, Nancy Frey, Linda M. Gojak, Sara Delano Moore, and William Mellman

Discover the right mathematics strategy to use at each learning phase so all students demonstrate more than a year's worth of learning per school year.

John Almarode, Douglas Fisher, Joseph Assof, Sara Delano Moore, Kateri Thunder, John Hattie, and Nancy Frey

Leverage the most effective teaching practices at the most effective time to meet the surface, deep, and transfer learning needs of every mathematics student.

Douglas Fisher, Nancy Frey, and John Hattie

Ensure students demonstrate more than a year's worth of learning during a school year by implementing the right literacy practice at the right moment.

Douglas Fisher, Nancy Frey, John Hattie, and Marisol Thayre

High-impact strategies to use for all you teach—all in one place. Deliver sustained, comprehensive literacy experiences to K–12 learners each day.

Nancy Frey, John Hattie, and Douglas Fisher

Imagine students who understand their educational goals and monitor their progress. This illuminating book focuses on self-assessment as a springboard for markedly higher levels of student achievement.

John Almarode, Douglas Fisher, Nancy Frey, and John Hattie

Inquiry, laboratory, project-based learning, discovery learning? The authors reveal that it's not which strategy is used, but when, and plot a vital K–12 framework for choosing the right approach at the right time.

Let us know what you think!

Did the information in this book resonate with you? We're hoping you'll continue to support this book's journey to reaching teachers and having the ultimate impact in the classroom. Here are a few ways you can do that:

>>> **JOIN** the conversation! Share your comments, participate in an online book study, or post a picture of yourself with the book on social media using **#VLClassroom**.

>>> **PROVIDE** your expert review of *Teaching Mathematics in the Visible Learning Classroom, Grades 3–5* on Amazon.

>>> **LEAD** or join a book study in your school or team to share ideas on how to bring the concepts presented in the book to life.

>>> **FOLLOW** our Corwin in the Classroom Facebook page and share your Visible Learning strategies in the mathematics classroom using **#VLClassroom**.

>>> **RECOMMEND** this book for your Professional Learning Community activities.

>>> **SUGGEST** this book to teacher educators.

Be sure to stay up-to-date on all things Corwin by following us on social media:
Facebook: www.facebook.com/corwinclassroom
Instagram: www.instagram.com/corwin_press, @corwin_press
Twitter: twitter.com/CorwinPress, @CorwinPress
Pinterest: www.pinterest.com/corwinpress/pins

www.corwin.com

CORWIN

A SAGE Publishing Company

Helping educators make the greatest impact

CORWIN HAS ONE MISSION: to enhance education through intentional professional learning.

We build long-term relationships with our authors, educators, clients, and associations who partner with us to develop and continuously improve the best evidence-based practices that establish and support lifelong learning.